SHEEP PRODUCTION: VOLUME ONE
BREEDING AND REPRODUCTION

RAY RICHARDS PUBLISHER
in association with
NEW ZEALAND INSTITUTE OF AGRICULTURAL SCIENCE

SHEEP PRODUCTION: VOLUME ONE

BREEDING AND REPRODUCTION

Edited by

G. A. WICKHAM and M. F. McDONALD

for the New Zealand Institute of Agricultural Science

RAY RICHARDS PUBLISHER

in association with the

NEW ZEALAND INSTITUTE OF AGRICULTURAL SCIENCE

First published 1982

NEW ZEALAND INSTITUTE OF AGRICULTURAL SCIENCE
Royal Society Building, 11 Turnbull Street, Wellington 1

RAY RICHARDS PUBLISHER
49 Aberdeen Road, Auckland 9

This book has been published with the assistance of the New Zealand Meat Producers Board and the New Zealand Wool Board.

Designed by Don Sinclair
Typeset by Linotype Service (P.N.) Ltd.
Printed by Colorcraft Ltd.

ABOUT THE SERIES

Technical information on aspects of sheep production tends to be widely scattered. Until now scientific and farming journals have been the main sources of information. However, the information seeker needed to have some idea where to find the appropriate articles and to have collected copies of a number of journals.

The New Zealand Institute of Agricultural Science has attempted to ease this problem by production of the present series of three volumes. The plan of the series was formulated in 1976 and involved allocating the important topics in sheep production and sheep products to a series of chapters. An author, technically competent in the area covered and skilled in authorship, was invited to produce each chapter. Authors were asked to summarise knowledge in a form that most non-scientists could readily understand. Other people with scientific and practical interest in the areas were asked to act as consultants, to check accuracy and relevance of the information, and the readability of the scripts. The editors were given the responsibility for coherence of the material and the uniformity of the style throughout.

The objective of the series is to provide a readily available source of knowledge for farmers, farm advisors, agricultural journalists, companies servicing the sheep industry, libraries and institutions, scientists and students. Sheepfarmers should find the volumes easy to read and the knowledge they acquire will help provide the basis for many farm management decisions. Many others on the fringe of sheep farming should also be able to benefit from a better understanding of the factors which help determine the nature of the sheep industry. Students should also find that this series goes a long way toward providing a textbook on aspects of sheep production.

While all the authors were domiciled in New Zealand at the time of writing and New Zealand systems of production figure largely, the material should also be of interest, value and application to people involved with the sheep industries of other countries.

Volume One: Seven chapters summarise information on the principles and practices of genetic improvement of sheep, while the final two chapters cover the reproductive processes of sheep and how these processes can be manipulated in farming situations.

Volume Two: Two chapters outline the basic processes of body and wool growth and how these may be changed. There are six chapters on sheep nutrition and three on aspects of sheep health.

Volume Three: The third volume covers some principles in managing sheep farms, together with chapters on the evaluation and marketing of meat, wool and sheep byproducts.

These three volumes provide a more complete coverage of the nature of sheep, their products and the systems used to exploit these than has been produced before.

CONTENTS

1

Introduction

A. D. H. Joblin
Ministry of Agriculture & Fisheries, Lincoln

NEW ZEALAND has the most highly developed sheep industry in the world. This statement is based on the fact that this country has carried 60 million sheep in recent years on nine million hectares of sown grassland and five million hectares of largely undeveloped tussock or *Danthonia* grassland. No major sheep producing nation approaches these figures in terms of carrying capacity per hectare. Wool production per animal in this country is also high, in comparison with that in overseas sheep producing nations. Over the period 1975 to 1978 New Zealand sheep averaged 5.2 kg of wool per head per annum. This was nearly double the average world production level of 2.7 kg per head. The sheep industry of this country is still growing in importance, with sheep numbers expected to rise in 1982 to over 70 million with wool yield per head at approximately 5.6 kg. This is still a long way below the potential that could be achieved from grassland farming.

This nation's export prosperity is more dependent on sheep production than on any other single industry. Over the four years from June 1976 to June 1979 inclusive, sheep meat and wool earned New Zealand 33% of its overseas income. This level of economic dependence on the sheep industry should have generated a profusion of authoritative books on all aspects of sheep breeding, management and product processing. In fact this has not happened.

One reason for this disturbing gap in the literature has been the extreme difficulty in gathering together the diversity of information that exists within the universities, research institutions, advisory services and in the accumulated experience of technically aware farmers. The information presented needs to be accurate, comprehensive and intelligible to the layman. No one man can expect to be expert in every facet of the industry and the smaller more specialised scientific societies would also have difficulty in covering the full range of disciplines that contribute to the welfare of the sheep industry.

It is therefore entirely appropriate that the New Zealand Institute of Agricultural Science, embracing every branch of agricultural science together with a full range of teaching, research, advisory and farming skills, should attempt this hitherto daunting task. The Institute is acutely aware of the need to get agricultural science applied in practice within the agricultural industry if science is to play its full part in lifting sheep farming productivity and with this the living standards of all New Zealanders. Knowledge is of little use locked up in academic notebooks, research files and esoteric scientific literature. It needs to be brought out into the open, critically examined and integrated to make up a complete and balanced account of the available scientific information pertaining to sheep farming.

In this book and others to follow on *Feeding, Growth and Health* and on *Farming Systems and Products*, the Institute has called on authors or groups of authors highly knowledgeable in their fields to prepare each of the chapters. These were then subject to scrutiny by "consultants" who were charged with checking the contributions not only for technical accuracy but also for their relevance to the industry and their readability to others less specialised in the topic under review. It then fell to the editors to draw the individual chapters together into a cohesive series of volumes on sheep production.

These books are aimed at the ever increasing numbers of farmers, who are actively seeking the most up-to-date information on sheep production under grazing management systems, as well as the many professional agriculturalists needing a ready work of reference on the sheep industry. The third group who should benefit are students at universities and technical institutes who have been poorly served by the lack of comprehensive reference books relevant to sheep production.

The series represents the culmination of thousands of hours of voluntary labour of scientists within and outside the Institute who have recognised the importance of the sheep industry to New Zealand, the contribution agricultural science can make to the productivity of that industry and the paramount importance of getting that knowledge applied within farming practice.

To all these contributors the Institute offers sincere thanks and commends their efforts to the people who can make their work worthwhile — the readers of these books.

2
Sheep Breeds in New Zealand

A. H. Carter and E. H. Cox
Ruakura Animal Research Station, Hamilton
Consultants: D. C. Dalton, K. C. Eastwood and G. A. Wickham

SUMMARY

The origins of New Zealand's present population of some 70 million sheep are traced through importations since the 1830's mainly from Britain and Australia, including breeds which have since disappeared. The establishment of breed societies and the development of derived local breeds are outlined.

Trends in breed distribution this century, summarized in terms of total ewes and rams and of ewes registered in flockbooks, are related to changes in farmland development, in relative economic returns from meat and wool and in market demands. The Romney had become the dominant breed by 1952 and comprised 78% of all ewes in 1967, the subsequent decline to 53% in 1977 being offset by growth of Perendale and Coopworth numbers to 28% of the total. The Southdown accounted for 93% of all Down rams in 1962, but only 55% in 1977 consequent on increasing popularity of the larger meat sire breeds.

Factors influencing the number and selection of rams of different breeds are discussed, with reference to breed society requirements. Experimental comparisons in New Zealand of the productive performance of different breeds and crosses are reviewed.

Improvement of existing breeds and successful development of new breeds have contributed greatly to the present efficiency of sheep production in New Zealand. Breed trends in the future will depend in part on the relative effectiveness of selection within breeds towards market-oriented production goals, in part on the demonstration and adoption of crossbreeding systems to enhance productivity and in part on the availability of superior new genetic material from overseas.

INTRODUCTION

New Zealand's economic development and prosperity owe much to sheep farming. Sheep products have contributed about half (two-thirds of this from wool) of total export earnings over the past century.[31] As in other countries, the evolution of sheep production and types of sheep has been determined in part by tradition, in part by availability of

11

animals and their adaptation to the environmental conditions, in part by market returns from alternative products.

Like other farm livestock, sheep are entirely exotic to New Zealand, the national flock deriving largely from importations from Britain and Australia. The dramatic growth in the sheep population is illustrated in Fig. 2.1.

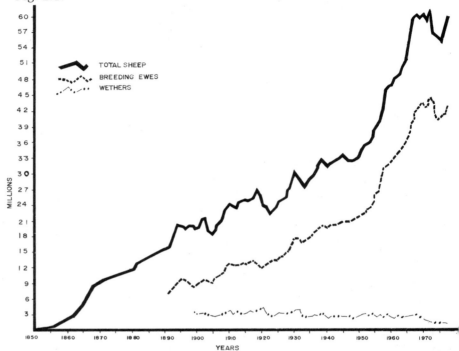

Fig. 2.1: Growth of the New Zealand sheep population from 1850 to 1977.

Numbers increased from under 1 million in 1855 to 70 million in 1981. The number and proportion of ewes in the population have increased steadily while wethers, traditionally run as wool producers in country not considered suitable for breeding stock, have declined numerically. This trend from wethers to ewes has been due to improved grazing conditions coupled with increased profitability of wool plus slaughter-lamb production systems relative to wool production solely.

Definition of Breed

A breed is a subgroup of a species possessing certain recognizable characteristics and maintained as a closed breeding population, historically in a single geographical area after which it is frequently named. In this chapter, *breeds* are defined as those subgroups which have been recognized through formation of a breed society or

12

registration of stock in an official flockbook. *Strains* will denote subdivisions of a breed not represented by separate breed associations.

Breed differences are partly the result of man's endeavours in selecting, transporting and crossbreeding to obtain the animals he desired and partly the result of natural processes of selection and "genetic drift". New breeds have originated and continue to originate as strains (e.g. Drysdale, Booroola Merino) or crosses (e.g. Corriedale, Coopworth) of existing breeds. Preoccupation with "breed purity" is thus of questionable value, with today's crossbred often becoming tomorrow's pure breed. Ryder[35] traces the probable evolution of British sheep breeds and relationships among them.

BREED INTRODUCTION AND ESTABLISHMENT IN NEW ZEALAND

Information has been derived from flock histories and entries recorded in official flockbooks, supplemented by some historical data.[37] [14] Only brief mention will be made of breed characteristics, which are well described elsewhere.[3] [23] [30] [34] The world's known sheep breeds have been classified according to functional use, colour, horns, origin and date of flockbook establishment.[28]

The New Zealand Sheep Breeders' Association (NZSBA) published the first national flockbook in 1895, representing 11 breeds. Separate flockbooks were issued annually for the North Island (1902-25) and South Island (1906-26) before re-amalgamation. Certain breed societies have established independent flockbooks, in some cases (Lincoln, Ryeland, Dorset Horn) only temporarily.

Table 2.1 lists the "effective" importations from which present New Zealand flocks derive, together with breeds which were introduced but which have since disappeared. Only the year of the first significant importation is shown although further introductions frequently followed. The dates in parentheses relate to importations resulting in only temporary establishment. Also tabulated are the years over which flockbook registrations have been recorded.

Base Breeds

Merinos constituted the first significant introduction of sheep, initially about 1814, with major importations from the 1830s onward. Australia was the predominant source with a few from America and Germany.[32] In consequence New Zealand Merinos are quite similar to those in Australia, with primary emphasis on wool rather than meat production. The importation of Merino rams from Australia has continued, with most being of strong-woolled (South Australian) and medium-woolled strains.[4]

TABLE 2.1: Historical Summary: Imported Pure Breeds

Type/Breed	Year Imported[a]	Flockbook Registrations NZSBA	Own Soc.
Base breeds established			
Finewool			
Merino (M)	(1814)1834	1895-	
Longwool (lw)			
English Leicester	(1843)1853	1895-	
Romney [Marsh][b] (R)	1853	1895-1959	1905-
Border Leicester (BL)	1859	1895-	
Lincoln (L)	(1840)1862	1895-	1912-74
Mediumwool			
Cheviot (Ch) ·	(1845)1857	1895-1910: 1936-65	1949-
Shortwool and Down			
Southdown (Sd)	(1842)1863	1895-1933	1926-
Shropshire [Down]	1864	1895-	
Ryeland	1901	1906-33: 1969-	1925-68
Suffolk [Down] (Sf)	1913	1914-	
Dorset Horn (DH)	(1897)(1925)1937	1906-11: 1926-27: 1937-	1939-53
Dorset Down (DD)	(1921)1947	1922: 1950-	
Hampshire [Down] (H)	(1861)(1881)1951	1895-1910: 1952-	
Breeds not permanently established			
Longwool			
Cotswold	(1863)	1895-1904	
Dartmoor	(1864)		
Wensleydale	(1894)(1920)(1952)	1895: 1952	
Roscommon	(1904)	1905-07	
South Devon	(1907)	1908-18	
German Whiteheaded			
Mutton	(1972)		
Carpetwool			
Scottish Blackface	(1908)		
Mediumwool			
Montadale	(1951)	1952	
Finnish Landrace	(1972)		
East Friesian	(1972)		
Shortwool and Down			
Norfoik Down	(1868)		
Tunis	(1900)		
Oxford Down	(1904)(1972)	1905-14	
Kerry Hill	(1937)	1938-50	
Wiltshire Horn	1973		

[a] Dates in parentheses relate to importations not resulting in permanent establishment of the breed.

[b] Name changed from Romney Marsh to New Zealand Romney in 1966.

Except for the Merino, all early sheep introductions were from Britain, comprising breeds familiar to the predominantly British settlers. Since 1952 however quarantine restrictions have precluded the introduction of sheep from sources other than Australia.

By 1865 the longwool breeds *Lincoln, English Leicester, Romney Marsh* (later New Zealand Romney) and *Border Leicester* together with the mediumwool *Cheviot* and the meat breeds *Southdown, Shropshire* and *Hampshire* were present. *Dorset Horn* and *Ryeland* sheep arrived about the turn of the century, followed by the *Suffolk* (1913) and *Dorset Down* (1921).

The initial importations did not always secure a permanent place for a breed. Both *Cheviot* and *Hampshire* registrations lapsed after 1910, the *Cheviot* being reinstated in 1936 supplemented by imports from Canada while today's *Hampshire* is based on new introductions from Britain and Australia since 1951. Final establishment of the *Dorset Horn* resulted from the importation of nearly 600 ewes from Australia during 1934-40, and of the *Dorset Down* from 105 ewes from England in the period 1947-1952. The *Suffolk*, represented by only one registered flock in 1938, was re-established through substantial importations from Australia in 1938-39.

Breeds Imported but Not Becoming Established

Among the breeds which were introduced but have since disappeared are several of potential present value.[5] From Britain came the longwoolled *Cotswold*, *Dartmoor*, *Wensleydale* and *South Devon*, the carpetwoolled *Scottish Blackface*, the *Oxford Down*, largest and heaviest-woolled of the Down breeds, and the black-nosed *Kerry Hill* related to the Shropshire and Ryeland; a reported *Norfolk Down* importation[37] may refer to the now-extinct Norfolk Horn or to its cross with the Southdown which became the Suffolk. Other introductions were, from Ireland, the longwoolled *Roscommon*, a forerunner of the Galway and from U.S.A., the *American Tunis*[37], derived from fat-tailed Tunisian sheep, and the *Montadale*, of Cheviot and Columbia (Lincoln x Rambouillet) ancestry. The fleece-shedding *Wiltshire Horn* was represented by a ram and four Poll Dorset cross ewes imported from Australia in 1974. Of all the above breeds only the Cotswold and Oxford Down became established in appreciable numbers.

In 1972 over 100 sheep of the prolific *Finnish Landrace*, the *East Friesian* milk sheep, the Cotswold-based *German Whiteheaded Mutton* and the *Oxford Down* breeds were imported from Britain and Ireland for experimental evaluation under stringent quarantine control.[7] Subsequent diagnosis of scrapie disease led to slaughter in 1978 of all these animals and their descendants.

Derived Breeds

Table 2.2. lists four breeds developed in Australia and subsequently imported to New Zealand. The *Poll Merino* and *Booroola Merino* originated as strains selected respectively for absence of horns and for multiple births. The *Polwarth* is an interbred Merino-Lincoln backcross or "comeback" (¾ Merino, ¼ Lincoln). The *Poll Dorset* resulted from crossing Dorset Horns with Corriedales or Ryelands and then repeatedly backcrossing with Dorset Horns while selecting against horns.

15

TABLE 2.2: Historical Summary: Derived Breeds and Strains

Type/Breed	Base breeds[a] Dam	Sire	Source	Year[b]	Flockbook NZSBA	Own Soc.
Finewool						
Polwarth	LxM	M	Aust.	(1932) 1957	1958-70	1973
Poll Merino	M	M	Aust.	1950	1950-	
Booroola Merino	M	M	Aust.	1972 1977[c]		
Carpetwool						
Drysdale	R[d]	R[d]	N.Z.	1971[c]		
Dual purpose (Meat + wool)						
Corriedale (Cd)	M	* lw	N.Z.	} 1903	1906[f]-47	1924-
Halfbred	M	* lw	N.Z.		1906[f]-	
Borderdale	Cd	* BL	N.Z.	1976	1977[f]-	
Perendale	R	* Ch	N.Z.	1960		1961-
Coopworth	R	BL	N.Z.	1968		1969-
Meat						
Poll Dorset	DH[e]	DH[e]	Aust.	1959	1959-	
South Suffolk	Sd	* Sf	N.Z.	1940	1940[f]-	
South Dorset Down	Sd	DD	N.Z.	1956		1958-
South Hampshire	Sd	* H	N.Z.	1974		1975-

*	Reciprocal cross permitted
a	For abbreviations refer Table 2.1; Cd = Corriedale.
b	Year imported (underlined), or Breed Society formed, or supplementary Flockbook approved by NZSBA.
c	Breed Society formed but no Flockbook yet published.
d	Romney "N-type" strain but may include some Cheviot blood.
e	Polled Dorset Horn strain derived from outcross to Corriedale or Ryeland.
f	Appendix established for crossbred sheep.

Several breeds have originated in New Zealand from endeavours to "fix" breed crosses combining desired attributes. In 1903 the NZSBA approved a flockbook appendix for "interbred halfbred sheep", ½ Merino, ½ British Longwool.[33] Full flockbook status was accorded the derived Corriedale and New Zealand Halfbred in 1916. The Lincoln and the English Leicester were the main longwool breeds contributing to the present *Corriedale* although the Otago property after which the breed is named used Romney-Merino crosses.[18] The *New Zealand Halfbred* is similar in appearance and origins to the Corriedale but does not require to be interbred. Of the rams used in registered Halfbred flocks in 1977, 30% were Halfbred, 66% were longwool (mainly English Leicester and Romney, mated with Merino ewes) and 4% Merino (mainly over Romney ewes). Commercial Halfbred ewe flocks generally use Halfbred rams but sometimes use Corriedales. The more recent *Borderdale* is an interbred Border Leicester-Corriedale cross which has been recognized by inclusion of a "grading-up" appendix in the NZSBA flockbook from 1977.

Cheviot-Romney crosses formed the basis of the *Perendale*, developed primarily to improve production from hard hill country. The *Coopworth* was derived from crossing Border Leicester rams with Romney ewes and subsequent interbreeding, with high performance levels being a requirement for flockbook registration.

Crossing the Southdown with larger Down breeds to improve growth rate has resulted in three "synthetic" meat sire breeds. The *South Dorset Down* is descended from the progeny of Southdown ewes and Dorset Down rams. Reciprocal crosses of the Southdown with the Suffolk and the Hampshire respectively formed the foundation of the *South Suffolk*, which attained full flockbook status in 1952, and the recently developed *South Hampshire*.

The *Drysdale* has been developed as a special purpose carpetwool sheep from a Romney strain carrying a genetic factor (N^d gene) for medullated or "hairy" fleeces, associated with growth of horns[22] (see Chapter 3 and photo 3.1). Animals may have some Cheviot ancestry, but only those homozygous for the N^d gene may be registered. Breed societies have not yet been founded for the *Tukidale* and Perendale-based *Carpetmaster* which have wool characteristics similar to the Drysdale but involving different genetic factors.[39]

BREED TRENDS IN THE SHEEP POPULATION

Few statistics exist on sheep breed numbers during the nineteenth century. The Merino initially adapted well to the natural tussock and produced a profitable wool clip. Forest clearing, cultivation and pasture establishment led to a dramatic expansion in sheep-carrying potential in a wide range of environmental conditions, many of which suited the British longwool breeds. The Lincoln, particularly favoured for its heavy fleece, was widely used to up-grade existing Merino and derived crossbred ewes. The Lincoln failed however to maintain its initial performance in the face of declining soil fertility, pasture reversion and internal parasites, resulting in a gradual swing towards use of the Romney which farmer experience established as better adapted to these conditions.

The advent of refrigeration in the 1880's marked a milestone in New Zealand's sheep history. While wool remained the major economic product, the added profitability of lamb meat production encouraged use of the "meatier" English Leicester and Border Leicester breeds and development of specialized fat lamb farming on improved country. By the turn of the century the national ewe flock comprised approximately one-third Merinos and Halfbreds, one-third Lincoln crosses (at least 50% Lincoln blood) and one-third Border Leicester, English Leicester and Romney crosses, with the Shropshire being the predominant fat lamb sire.

2.1

2.2

Photo 2.1: A New Zealand Romney ram. This breed which originally evolved in a wet, exposed environment in Southern England has been asked to cope with a wide range of environments, from flat to steep hills, in New Zealand. (Opposite top)

Photo 2.2: The Corriedale, the first synthetic breed developed in New Zealand, evolved from British Longwool x Merino crosses. Now found in many areas of the world, it combined high fleece weights with good quality wool and reasonable lamb production. (Opposite bottom)

Photo 2.3: The Perendale which has tended to replace Romneys on much less productive, wet hill country in New Zealand. While wool weights are not as great as that of the Romneys, the wool is more bulky and more lambs are produced. (Below)

2.3

2.4

2.5

Photo 2.4: The Coopworth which is difficult to distinguish from the Perendale by laymen, but the sheep are bigger with coarser and heavier fleeces. They have tended to replace Romneys on more productive land when farmers wish to produce more lambs. (Opposite top)

Photo 2.5: The Southdown. This was the absolutely dominant meat breed in New Zealand when the market demanded small, fatty carcasses. The ram illustrated is of the less-compact modern type. (Opposite bottom)

Photo 2.6: The Suffolk which is a larger, leaner meat breed is, with other breeds, now providing stiffer competition for the Southdown. (Below)

2.6

Sources of Breed Distribution Data

Although breed society flockbooks provide accurate records of numbers of registered ewes, the breed composition of the national sheep flock can be estimated only approximately from periodic censuses which itemize the main breeds only.

Commencing with the Sheep Act of 1878, all sheep owners have been required to furnish an annual return of numbers of sheep in specified age, sex and "breed" (every fifth year only since 1952) categories. In early years sheep were classified into those of (listed) "distinctive" breeds; Merino crosses or "flock sheep"; and "crossbreds", much the largest group, which included longwool crosses and breeds not otherwise specified. The definition of "sheep of a distinctive breed" was broadened in 1952 to include many previously designated as "crossbreds". From 1917 to 1972 sheep in stud flocks were classified separately, this category being extended in 1977 to include registered (stud) and unregistered ram breeding flocks.

A design fault in the census forms since 1952 has caused bias through some ewes being classified according to the breed of ram to which they were mated rather than to their own breed. Adjusted ewe numbers for the Down, Border Leicester and Cheviot breeds have been calculated from corresponding ewe hoggets by assuming a true replacement rate of 0.3 and the remaining ewes, presumed incorrectly ascribed to these "crossing" breeds and comprising up to 2% of all ewes, distributed proportionately among all breeds. Sheep classified as of "other breeds" have been apportioned where appropriate on the basis of known importations and developments in registered flocks and in relation to previous or subsequent breed survey data.

Traditionally the rams used by commercial farmers have been bred in flocks registered with breed societies. Thus periodic analysis of the numbers of flockbook or "stud" ewes of the different breeds provides useful information on the potential for breed expansion or breed change through crossbreeding, as well as on the intensity of flock ram selection. In recent years however, increasing numbers of commercial rams have been bred in non-registered flocks, often associated with stud flocks or group breeding schemes, but relevant statistical data are unavailable.

Breed Composition of the National Flock Since 1902

Total numbers of ewes two-tooth (18 months) and older and their percentage distribution among the main breeds are summarized in Table 2.3 at intervals from 1902.

Figures for 1977 relate to ewes "put to ram" and hence exclude unmated ewes (which in three previous surveys comprised less than 1% of all ewes). The overall percentages of flockbook ewes are based on census returns except for the first and last entries which relate to actual flockbook registrations in 1901 and 1977 respectively.

Corresponding breed distribution data for rams (two-tooth and older) are presented in Table 2.4, together with the overall ratios (%) of rams to ewes.

Changing breed patterns are clearly illustrated in Fig. 2.2 (ewes) and Fig. 2.3 (rams); "new breeds" refer to Perendale, Coopworth and Drysdale. "Crossbreds", which constituted the bulk of the national ewe flock up to 1952, derived largely from "grading up" of Merino or

TABLE 2.3: New Zealand Ewe Flock 1902-77: Percentage Breed Distribution.

Ewes two-tooth and older up to 1972: ewes put to ram in 1977. Figures in (   ) are estimated percentages for breeds included in "Other ". + Denotes under 0.05% of total

Breed	1902	1922	1940	1952	1962	1967	1972	1977*
TOTAL ('000)	11035	13456	20693	23414	34494	41787	44578	42782
Flockbook ewes (%)	0.9	1.2	1.3	1.4	1.5	1.3	1.3	1.2
Merino	0.3	3.0	2.1	1.6	1.2	1.0	1.2	1.1
Halfbred	11.9[a]	4.6	6.7	7.3	4.1	3.6	4.0	4.2
Corriedale	—	2.5	4.0	4.2	5.3	6.1	7.8	8.8
Merino-based	12.2	10.1	12.8	13.1	10.6	10.7	13.0	14.1
Romney	0.4	10.9	13.2	64.6[a]	75.0	78.1	68.7	52.6
Crossbred	81.6[a]	76.7	73.1	21.2[a]	13.1	9.1	13.8[a]	1.8[b]
Other (ewe)	—	(0.1)	—	—	(0.2)[c]	(0.7)[c]	(2.2)[c]	29.1[d]
Romney-based	86.5	87.7	86.3	85.8	88.3	87.9	84.7	83.5
Lincoln	0.6	0.8	+	+	(+)	(+)	(+)	+
English Leicester	0.3	0.6	0.1	+	(+)	(+)	(+)	+
Border Leicester	0.2	0.5	0.1	0.1	0.1	0.6	1.3	1.8
Cheviot	(+)	(+)	(+)	0.1	0.2	(0.2)	(0.1)	0.1
Crossing breeds	1.1	1.9	0.3	0.2	0.3	0.8	1.5	1.9
Southdown	+	0.2	0.7	0.8	0.7	0.4	0.4	0.2
Other Down	(0.2)	(0.1)	(+)	(+)	(0.1)	0.2	0.4	0.3
Down breeds	0.2	0.3	0.7	0.9	0.8	0.6	0.8	0.5
Other	0.2[e]	0.2[e]	+[ef]	+[fg]	0.3[f]	0.9[h]	2.3[h]	0.1

*	Unofficial calculations based on unpublished sheep return data
a	See text.
b	Excludes Border Leicester-Romney and Cheviot-Romney cross; see note d.
c	Presumed mainly Perendale.
d	Includes Perendale plus Cheviot-Romney cross, 15.2%; Coopworth plus Border Leicester-Romney cross, 13.0%; Drysdale, 0.8%.
e	Includes Shropshire, 0.14% in 1902, under 0.05% in 1922 and 1940.
f	Includes Ryeland, under 0.05%.
g	Includes Dorset Horn, under 0.05%.
h	Excludes "Other Down".

Merino-cross ewes by use initially of Lincoln and to a lesser extent of Romney, English Leicester and Border Leicester rams. The replacement as the dominant sire breed of the Lincoln by the Romney, found to be more resistant to footrot and parasite problems, was clearly established by 1922. Continued use of the Romney in the grading up process led to its emergence as the predominant ewe breed in the national flock when "crossbred" ewes were largely re-classified in 1952 according to their

TABLE 2.4: New Zealand Ram Flock 1902-77: Percentage Breed Distribution
Rams two-tooth and older.

Figures in ( ) are estimated percentages for breeds included in "Other".
+ Denotes under 0.05% of total

Breed	1902	1922	1940	1952	1962	1967	1972	1977*
TOTAL ('000)	218.0	322.1	541.1	632.1	899.7	980.4	1029.1	876.6
Rams/Ewes (%)	2.0	2.3	2.6	2.7	2.6	2.3	2.3	2.1
Merino	13.8	4.8	2.7	2.2	1.1	1.0	1.3	1.3
Halfbred	(4.6)[a]	5.5	5.1	4.6	2.6	2.7	3.3	3.4
Corriedale	—	5.7	5.2	3.9	3.2	4.2	5.4	6.5
Merino-based	18.4	16.0	13.0	10.7	6.9	7.9	10.0	11.2
Romney	17.8	54.0	52.4	54.8	56.3	67.1	56.0	44.7
Crossbred	—	—	(0.7)	5.0	4.6	4.3	9.4[a]	1.6[b]
Other (ewe)	—	—	—	—	(0.1)[c]	(0.2)[c]	(1.9)[c]	22.0[d]
Romney-based	17.8	54.0	53.1	59.8	61.0	71.6	67.3	68.3
Lincoln	30.2	6.1	0.3	0.2	(0.1)	(+)	(+)	(+)
English Leicester	14.4	7.4	2.5	0.3	0.3	(0.2)	(0.1)	0.1
Border Leicester	11.0	8.2	2.9	1.6	2.2	1.9	3.3	3.7
Cheviot	(1.0)	(0.5)	(0.2)	0.4	0.7	(0.7)	(0.7)	0.8
Crossing breeds	56.6	22.2	5.9	2.5	3.3	2.8	4.1	4.6
Southdown	1.6	6.5	26.7	26.3	26.9	14.7	12.0	8.7
Shropshire	5.4	1.0	0.5	(+)	(+)	(+)	(+)	+
Ryeland	—	(+)	0.7	0.4	0.3	(+)	(+)	0.1
Other Down	(0.2)	(0.3)	(0.1)	(0.3)	(1.6)	3.0	6.6	7.1[e]
Down breeds	7.2	7.8	28.0	27.0	28.8	17.7	18.6	15.9
Other	5.8	0.8	1.0	0.3[f]	1.8	1.1[g]	2.7[g]	0.2

*	Unofficial calculations based on unpublished Sheep Return data, 1977.
a	See text.
b	Excludes Border Leicester-Romney and Cheviot-Romney cross.
c	Presumed mainly Perendale.
d	Includes Perendale plus Cheviot-Romney cross, 12.9%, Coopworth plus Border Leicester-Romney cross, 8.2%; Drysdale, 0.6%.
e	Includes Suffolk, 0.9%; South Suffolk, 2.0%; Dorset Down, 1.2%; South Dorset Down, 1.2%; Poll Dorset plus Dorset Horn, 1.1%; Hampshire 0.5%.
f	Includes Dorset Horn, 0.2%
g	Excludes "Other Down".

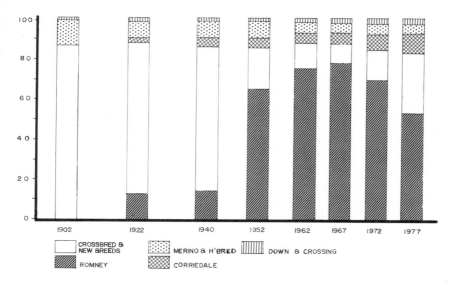

Fig. 2.2: Percentage of the New Zealand ewe population represented by various breed categories are indicated by the national sheep returns.

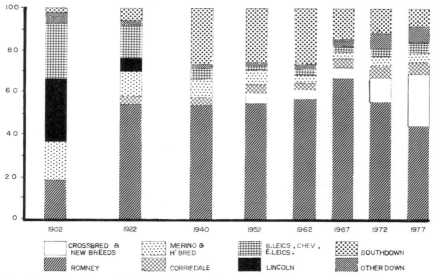

Fig. 2.3: Percentage of the New Zealand ram population represented by various breed categories.

distinctive breed type. The popularity of the Romney breed reached a peak around 1967 when it accounted for 78% of all ewes and 67% of rams. Despite its subsequent numerical decline the Romney, with 22.3 million ewes on 1977, remains much the most numerous sheep breed in the country.

In the last decade numbers of the two Romney-based synthetic breeds, Perendale and Coopworth, have increased to 15% and 13% respectively of total ewes in 1977, mainly at the expense of the Romney to which they proved superior in productivity under some conditions. Although crossbreds were not itemized as to parent breeds prior to 1977, a large proportion of the crossbred ewes and rams in 1972 would have been Cheviot x Romney or Border Leicester x Romney crosses. The dominant role of the Romney breed throughout this century is clear. Over the whole period Romneys and crossbreds, largely Romney-based, have accounted for about 85% of all ewes. Approximately 70% of all genes in present ewe flocks are estimated to be from the Romney.

Merino and Halfbred ewe numbers have remained fairly stable since 1940, although their proportion of total ewes has declined. By contrast, Corriedales have increased steadily in both relative and absolute terms to reach 3.7 million ewes, 8.8% of the total, in 1977.

For the Merino-based and Romney-based breeds, the relative distribution of rams broadly parallels that of ewes, differences between the two sexes resulting from use of other breeds of ram for crossbreeding.

The dual-purpose crossing breeds and the Down breeds contribute less than 2% and 1% respectively to the national ewe flock, a substantial proportion of these being in stud flocks. The importance of these breeds is seen rather in their contribution to the national ram flock (Table 2.4, Fig. 2.3). Among the Longwool breeds, the most striking change has been the virtual eclipse of the Lincoln and English Leicester which together accounted for 45% of all rams in 1902. The proportion of Border Leicester rams also declined until 1952 but then increased in response to growing demand for the breed to sire crossbred ewes or export lambs.

Usage of specialized meat sire breeds has closely reflected trends in production patterns and overseas market demands. The Shropshire, initially favoured for mutton and heavy-weight lamb exports, was rapidly replaced early this century by the Southdown, well suited for early drafting to meet a developing lighter-carcass market. By 1962, the Southdown accounted for 27% of all rams and approximately 93% of all Down rams. Changing export demands towards heavier leaner carcasses were responsible in turn for the recent significant swing to the larger meat breeds — notably Suffolk, South Suffolk, Poll Dorset or Dorset Horn, Dorset Down and South Dorset Down — at the expense of the Southdown, whose share of all Down rams dropped to 55% by 1977.

The ratio of rams to ewes is shown at the top of Table 2.4 for all breeds combined. The decline from 1952 (2.7%) to 1977 (2.1%) largely reflects increasing numbers of ewes mated by each ram, since research and farmer experience showed that this was a sound practice. Variation in the ram/ewe ratio among breeds is primarily associated with differences in usage for crossbreeding.

TABLE 2.5: Numbers of Registered Flocks (N) and Breeding Ewes (E)
1901-77
From Flockbook entries unless otherwise indicated: Ewes 2th and older

Breed	1901		1921		1940		1960		1977	
	N	E	N	E	N	E	N	E	N	E
TOTAL	478	102963	1282	183429	2321	297469	4087	508979	3573	500006
Merino	26	9358	33	11863	48	10989	46	11315	45	9787
Poll Merino							3	245	12	1036
Halfbred[a]			10	5874	37	6457	51	8518	30	6784
				(510)		(4058)		(5805)		(4304)
Corriedale			75	19850	193[b]	30643	164	28657	179	41982
Romney	59	20075	561[bw]	80355	769[bx]	132077	1258	261791	723[c]	168358
Perendale							24[cz]	1894	284[c]	60301
Coopworth									152[c]	34420
Lincoln	122	28793	182[by]	21871	18[b]	1649	14[b]	1821	12	1096
E. Leicester	113	19141	108	14575	100	11711	33	2349	21	2043
B. Leicester	67	12836	143	13445	128	9963	347	18573	320	34801
Cheviot	4	2116			5	268	101[b]	6102	76[c]	7281
Southdown	19	2299	138	13366	862	86289	1691	152439	455	41305
Shropshire	64	7932	20	1634	18	905	10	288	2	100
Ryeland			9	525	105	5551	39[z]	2008	26	969
Suffolk			3	71	7	170	54	1963	205	11097
South Suffolk					1	150	129	6286	236	16899
Dorset Horn					29[bd]	632	45	2572	15	837
Poll Dorset							6	276	207	19249
Dorset Down							22	536	170	10873
S. Dorset Down							36[z]	1107	187	14528
Hampshire	3	305					13	222	126	6129
Other	1[e]	108			1[f]	15	1[g]	17	90[hc]	10131

[a] Ewe totals for Halfbred flocks include purebred Merinos shown in parenthesis.

[b] Entries combined from NZSBA and separate breed Flockbooks.

[c] Figures provided by Breed Association(s).

[d] Entered in 1940 (267) or 1939 (61) Flockbooks plus additional ewes imported before 1940 (141) or imported in lamb in 1940 (163).

[e] Cotswold.

[f] Kerry Hill.

[g] Polwarth.

[h] Drysdale 37 flocks, 5000 ewes; Borderdale 23 flocks, 2715 ewes; South Hampshire 30 flocks, 2416 ewes.

[w] Returns for 1920.

[x] Returns for 1939.

[y] Returns for 1919.

[z] Returns for 1961.

Breed Distribution Trends Among Registered Flocks

Table 2.5 shows the numbers of registered flocks and breeding ewes recorded for each breed at intervals from 1901 to 1977. Entries in flockbook appendices have been included. After a five-fold expansion from 1901 to 1960, the total of flockbook ewes has since remained relatively constant. As a proportion of total ewes in the population, stud ewes increased from 0.9% in 1901 to 1.5% in 1960 and then declined to 1.2% in 1977.

The main trends closely follow those in the national ewe and ram flock, namely a dramatic increase in the Romney (to 1969), Southdown (to 1960) and more recently Perendale, Coopworth and the larger meat sire breeds; a steady expansion of the Corriedale, Border Leicester and Cheviot; and a sharp decline in Lincoln, English Leicester, Shropshire and Ryeland registrations. The Poll Dorset has recently expanded at the expense of Dorset Horn flocks "grading up" through use of Poll Dorset rams.

Average flock size has varied to only a small degree over the years but differs widely among breeds. Stud flocks of the Merino and dual-purpose breeds average about 200 ewes, of the dual-purpose crossing breeds and the Southdown about 100 ewes and of the other Down breeds about 60 ewes.

The ratios of stud ewes to total ewes, shown for all breeds combined in Table 2.3, vary widely among breed groups and to some extent from year to year. Before 1950 about 5% of "distinctive" Romney ewes were in stud flocks, the proportion then dropping to about 1% in consequence of the reclassification of many "crossbred" ewes as Romneys. Stud ewes have fallen steadily as a proportion of all Corriedales, from 6% in 1922 to 1% in 1977. Registrations of Merino ewes have varied between 5% and 3% (but only 2% on latest figures). The proportion of Border Leicester and English Leicester ewes entered in flockbooks averaged 40% up to 1962 but has since decreased sharply to under 1%. Registered ewes have comprised on average 55% of total Southdown ewes.

STRUCTURE OF RAM BREEDING INDUSTRY

The genetic production potential of commercial flocks is determined almost entirely by the intensity and the accuracy of selection towards production-oriented goals in ram breeding flocks and by the proportion of rams bred which are actually used. Two important factors which condition the effectiveness of within-flock selection are flock size, affecting selection intensity, and the extent of performance recording, influencing the accuracy of selection.

Table 2.6 summarizes, by breeds, information for 1977 on rams, ewes (A) in ram breeding flocks and flockbook ewes (F), relative to total ewes (C). Comparison of the second (A/C%) and fourth (F/C%) columns shows that the proportion of all ewes in ram breeding flocks which were entered in flockbooks was approximately 50% for each of the Merino-based and Romney-based breeds except for Drysdale and "other", 75% for the dual-purpose crossing breeds and 95% for the Down breeds. Ram/ewe ratios (R/C%) are naturally highest for the Down breeds, lowest for those breeds used widely as export lamb dams and intermediate for the dual-purpose crossing breeds; a ten-fold decline since 1962 in the value for the Border Leicester largely reflects the sharp increase in the number of ewes of this breed (see Table 2.3). The higher ratio for the Merino than for the other main ewe breeds results from its limited use as a prime lamb dam.

TABLE 2.6: Breed Distribution 1977: Census, Flockbook and Sheeplan Data N = no. flocks; E = average no. ewes per flock

Breed	Census[a]			Flockbook[b]		Sheeplan[c]		
	C('000)	$\frac{A}{C}$(%)	$\frac{R}{C}$(%)	$\frac{F}{C}$(%)	Ē	N	Ē	$\frac{S}{F}$(%)
TOTAL	42373*	2.3	2.1	1.2	140	871	257	45
Merino	482	5.6	2.4	2.2[d]	190	5	170	8
Halfbred	1769	0.9	1.7	0.4	226	1	120	2
Corriedale	3733	2.0	1.5	1.1	235	41	370	36
Borderdale	—[e]			(2715)[f]	118	17	120	75
Polwarth	2	49.0	11.8	—	—	2	188	16
Merino-based	5986	2.0	1.6	1.0	216	66	278	29
Romney	22297	1.9	1.8	0.8	233	225	334	45
Perendale	6460[g]	2.4	1.8	0.9	212	148	302	74
Coopworth	5505[h]	1.4	1.3	0.7	226	177	288	148
Drysdale	328	2.0	1.7	1.5	135	17	269	92
Other	778[k]	1.5	2.0	—	—	7[l]	144	—
Romney-based	35368	1.9	1.7	0.8	224	574	308	66
Lincoln	2*	55.4	9.6	61.1	91	—		
English Leicester	24	10.3	5.1	8.4	97	—		
Border Leicester	743*	6.3	4.4	4.7	109	48	134	18
Cheviot	40*	21.1	17.2	18.0	96	19	139	36
Crossing breeds	809	7.3	5.0	5.6	105	67	136	20
Southdown	77*	61	99	54	91	25	155	9
Shropshire	—	68	73	40	50	—		
Ryeland	1*	100	59	98	37	1	67	7
Suffolk	20*	57	38	55	54	20	70	13
South Suffolk	28*	64	64	61	72	14	109	9
Dorset Horn	3*	28	45	25	56	—		
Poll Dorset	35*	47	23	55	93	24	172	21
Dorset Down	16*	77	69	68	64	31	103	29
South Dorset Down	20*	76	50	71	78	39	130	35
Hampshire	6*	93	73	98	49	9	88	13
South Hampshire	3*	98	37	85	81	1	65	3
Down breeds	210	62.4	66	59.2	75	164	123	16

* Ewe numbers adjusted (see text).

[a] Unofficial calculations based on unpublished Sheep Return data, 1977. C = total breeding ewes; A = ewes in Stud and Ram Breeding flocks; R = total rams 2-tooth and older.

[b] F = ewes entered in 1977 Flockbooks — see Table 2.5.

[c] S = ewes performance recorded under Sheeplan in 1977 (36).

[d] Includes Poll Merino (12 flocks averaging 86 ewes each).

[e] Borderdale not identified in census — included in "Other".

[f] Ewes entered in NZSBA Flockbook Appendix, 1977.

[g] Includes Cheviot-Romney cross.

[h] Includes Border Leicester-Romney cross.

[k] Miscellaneous crossbreds, 749 thousand; "other breeds", 29.

[l] Tukidale, 2 flocks; Carpetmaster, 1; Cheviot-Corriedale, 1; Romney-Corriedale, 1; and "Pigmented", 2.

The New Zealand National Flock Recording Scheme (Sheeplan) publishes an annual list of the flocks using the scheme. Data for 1977[36] (table 2.6) show that these flocks have more sheep than the average registered flock.

The final column (S/F%) expresses Sheeplan recorded ewes, which include some in non-registered ram breeding flocks and group breeding schemes, as a percentage of ewes of that breed entered in flockbooks. The proportions would be approximately halved for the dual-purpose breeds if based on ewes in *all* ram breeding flocks. The high Sheeplan participation for the Coopworth, Drysdale, Perendale and Borderdale breeds reflects the obligation for performance recording as a basis for registration. For most Down breeds, as for the Merino and Halfbred, only a small proportion of registered flocks participate in Sheeplan.

Assuming a mating (ram/ewe) ratio of 2.1% and an average ram working life of three years, annual ram replacements should total 0.7% of the ewe population. An effective two-tooth ram generating rate of 40% (allowing for minimum culling) would then require 1.75% of all ewes to be in ram breeding flocks. Reference to Table 2.3 (second line) suggests firstly that in the past when most flock rams were supplied by registered stud flocks virtually all the rams bred were in fact used, and secondly that an increasing proportion of commercial rams is now coming from non-registered flocks.

Breed differences in the ratio of rams to "ram breeding" ewes provide a measure of differences in the proportion of rams bred which are actually used for breeding. Excluding Halfbred flocks, many of which use Merino or longwool rams, the ratio is highest for the Southdown (1.6) and ranges from 0.6 to 1.0 for other breeds, the overall average being 0.9. Assuming again an average ram working life of three years, it appears therefore that most rams produced in Southdown ram breeding flocks are used, the proportion varying from about 50% to 70% among other breeds.

The stud breeder has historically been the principal agent in livestock improvement, with many present breeds testifying to his success. Breed societies have evolved to serve the interests of their members, initially through promotion and preservation of those distinctive features of the breed regarded as desirable, although betterment of the breed has always been an avowed objective. In recent times however, emphasis on "breed purity", regulated through closed flockbooks, and on conformity to "breed type" has frequently tended to limit improvement in terms of present-day production requirements.

To a varying extent, breed societies require or encourage practices designed to enhance genetic improvement. Minimum culling levels among ewe lambs are prescribed by the NZSBA (10% up to two-tooth age) and by the Corriedale (20% as two-tooths), Perendale (50% by fourth lambing) and Coopworth (40% to four-tooth mating) Breed Associations. Registered flocks for the Southdown breed must com-

30

prise at least 10 ewes, for the Romney 5 (previously 15 or 25), for the Perendale 50 (previously 25) and for the Coopworth 50 ewes. In contrast to the older-established breeds, flockbooks for the Perendale and Coopworth are not closed, certification of approved performance standards as well as pedigree being a pre-requisite for entry. Recording of performance under Sheeplan is obligatory for continued registration of Coopworth flocks. Regulations for the Drysdale, Borderdale and Booroola Merino breeds parallel those for the Perendale or Coopworth.

BREED CHARACTERISTICS AND COMPARATIVE PERFORMANCE

All New Zealand's sheep are basically white-woolled although recessive genes for wool pigmentation occur at low frequency in many breeds. Recent interest in the breeding of sheep with dark-coloured fleeces for use in home spinning is testified by formation in 1976 of the New Zealand Black and Coloured Sheep Breeders' Association. The Merino and its derivatives produce the finest wool, followed in order by the Down breeds, Cheviot, Romney, Border Leicester, English Leicester and Lincoln.[23] Except for the Merino and Dorset Horn, for which hornless strains now exist, and the newly developed carpetwool types, all local sheep are naturally polled. Black or brown skin pigmentation, frequently accompanied by presence of some dark wool fibres, is characteristic of the Suffolk, Dorset Down, Hampshire and Shropshire and to a lesser extent the Southdown. The Border Leicester, Cheviot, Suffolk and Dorset (Horn or Poll) have open (wool-free) faces and lower limbs, the extent of wool cover showing some variation within and among the other breeds. The physical attributes of crosses, and of derived "synthetics", are usually intermediate to those of the foundation purebreds.

Assessment of breeds must obviously be related to their functional use, market requirements and the conditions under which they are farmed. In New Zealand's predominantly two-tier sheep industry structure, breeding flocks maintained on hill country and poorer pastures provide surplus ewes as dams for export lamb farming on more productive land. For the former enterprise wool production traits of the sheep are more important relative to fertility and meat production traits, than is the case on more productive land. Higher returns for meat relative to wool coupled with improved pastures and management on hill farms, enhancing the survival rate of twins and permitting rearing of lambs to slaughter, have increased the economic importance of fertility as against wool production. Down breeds must be judged primarily on lamb survival, growth rate and carcass merit in relation to meat buyers demands. Present and future market requirements are critical also in evaluating and comparing speciality wool breeds such as the Merino and Drysdale.

Factual information on the relative productive merits of breeds is limited. On the one hand individual performance is seldom recorded except in experimental and some ram breeding flocks; on the other, different breeds tend to be maintained on different properties and frequently in different regions or classes of country, precluding meaningful performance comparisons. Recourse is therefore necessary to results from experiments in which animals of different breeds or crosses are compared under similar conditions.

Relevant experiments fall into three broad categories: comparison of representative sample flocks of the different breeds; progeny test evaluation of sire samples, mated to a common dam breed; and crossbreeding trials designed to measure genetic parameters of breeds and crosses. When crosses are involved care must be taken that hybrid vigour effects are not confused with true breed differences.[6] Reciprocal crossing studies among the Romney, Border Leicester, Cheviot and Merino breeds at Woodlands,* and the Romney, Dorset and Corriedale at Templeton show performance superiority of the first-cross ewes over the average of the straightbreds amounting to 29% at Woodlands and 10% at Templeton for lamb weaning percentage, compared with about 4-9% for live weight and fleece weight.[10] The advantage in crossbred performance due to heterosis is expected to halve when the first crosses are interbred (without selection) but to remain stable with further interbreeding (see Chapter 5). This is borne out by experience with the Cheviot-Romney cross at Massey University,[23] with the Border Leicester-Romney cross at Lincoln College[15] and at Whatawhata,[26] and with the Merino-Romney cross at Tokanui[8].

In the most extensive breed comparison study yet undertaken, 15,000 Romney ewes were mated with 435 rams of 14 breeds, all progeny being slaughtered for carcass assessment at 3-6 months of age. Sire breed rankings relative to the Southdown are summarized in Table 2.7.[9]

Important differences occurred among sire breeds in progeny survival, for which the Southdown excelled. Progeny of longwool sires clipped more wool at postweaning shearing than of the Down breeds, with the Cheviot, Dorset and Ryeland intermediate. Breed differences in progeny liveweights broadly followed those of the (two-tooth) sires themselves except that, relative to ram weights, Dorset, Cheviot and Southdown crosses showed rapid early growth with the reverse being the case for the "slow-maturing" English Leicester, Border Leicester and Romney breeds. Slaughter live weight provided a satisfactory criterion of profitability, the lower carcass:live-weight ratio of the longwool than the Down crosses being compensated by their greater wool pull. Observed breed differences in conformation, composition

---

*Woodlands, Templeton, Ruakura, Invermay, Tara Hills, Tokanui and Crater will refer to the MAF Research Division's Centres or Stations at these locations.

TABLE 2.7: Comparison of Sire Breeds for Lamb Production: Ruakura, 1963-73[9]

Sire Breed	Ram (2-tooth) liveweight	Progeny (from Romney ewes)				
		Survival rate	Fleece weight	120-day weight	Carcass weight	Net Merit
Southdown	68kg = 100	89.3% = 100	0.83kg = 100	26.0kg = 100	13.0kg = 100	100
Dorset (Horn or Poll)	106	96	110	107	108	103
Suffolk	125	94	102	108	109	102
Dorset Down	117	97	100	106	106	102
South Suffolk	113	97	101	105	105	102
South Dorset Down	115	96	97	103	104	101
Border Leicester	130	96	124	106	104	101
Hampshire	120	91	101	108	105	98
English Leicester	122	96	127	101	98	97
Cheviot	97	95	103	101	100	96
Ryeland	105	94	110	98	96	92
Lincoln	99	94	145	98	92	92
Merino	85	97	115	93	87	90
Romney	105	90	128	93	86	84

and commercial grades of the carcasses were of minor economic importance under the heavy stocking rates and hence comparatively light slaughter weights in the trial. The net merit figures in Table 2.7, calculated as the product of survival rate and 120-day progeny live weights, reflect economic return per ewe mated. The poor ranking of the (straightbred) Romney is presumed due in part to the absence of heterosis, all other breeds siring crossbred progeny. Within any breed, wide differences existed between progeny growth rates of the best and poorest sires, emphasizing the importance of sound selection of rams and of adequate genetic sampling in breed comparisons.

Other experiments [11] [16] [17] and meat industry surveys [21] have broadly substantiated the sire breed rankings for growth rate and carcass weight in the above trials.

Three trials [20] compared representative breeding flocks of Romney, Perendale and Coopworth on hill country (Whatawhata) and at two (Invermay) or three (Ruakura) stocking rates on lowland farms. Romney ewes had the lowest live weights and weaned the lightest lambs but produced the heaviest fleeces; the Perendales were lighter, with substantially lower fleece weights, than the Coopworths, but both breeds weaned appreciably more lambs than the Romney. It was concluded that economic productivity of the two "synthetic" breeds exceeds that of the Romney under all conditions, with the Coopworth being superior to the Perendale only in very good conditions.

In experiments at Woodlands[10] purebred Border Leicesters were heavier, clipped more, but coarser, wool and weaned more lambs than Cheviots. Romneys were intermediate for body weight, highest for wool production and poorest for number of lambs weaned. Similar rankings applied to progeny sired by these three breeds both at Woodlands and at Crater,[12] except that the larger Border Leicester-sired progeny clipped more wool than those by Romney sires.

The Merino, included in the Woodlands trial, had substantially lower live weight and weaning percentage than the other three breeds, but approached the Romney in fleece weight. In comparisons with the Corriedale and Perendale[38] and with the Halfbred[27], the lighter Merino was concluded to be the more profitable breed in returns per ha. from wool plus lamb despite sometimes lower lambing percentage. At Tokanui in the Waikato,[8] Merino ewes were 15% lighter than control Romney ewes but weaned similar numbers of lambs over a four-year period; the lighter Merino hoggets also clipped more wool than Romney hoggets. Merino-Romney cross ewes were far superior to either pure breed in lambing performance.

At Templeton[10] the Romney and Corriedale breeds were comparable in terms of ewe live weight, wool production and number of lambs born, with Corriedale lambs showing better survival to weaning; Dorset (Horn or Poll) sheep were heavier, produced only 60% as much wool, but had a substantially higher lambing rate although this superiority was reduced by higher lamb losses to weaning. Superior milk production in the Dorset ewe[25] was reflected in high progeny weaning weights. Lambs from Dorset and Dorset-cross ewes grew faster and produced heavier carcasses than those from Romney and Corriedale dams.[24]

Dorset-Romney crosses at Whatawhata ranked above the Romney and below the Coopworth and Perendale for live weight and for weaning percentage but below these three breeds for ewe and hogget fleece weights.[19] By contrast, Dorset x Romney ewes at Crater were markedly superior to Border Leicester x Romney and Cheviot x Romney for lamb production (the straight Romney being poorest in this trait). The Dorset and Cheviot crosses were similar in wool production but both were inferior to the Romney and the Border Leicester x Romney. The Dorset and to a lesser extent the Merino are characterized by a longer breeding season (shorter anoestrum) than other local breeds. This could be of value for early or "out of season" lambing.

Although early studies suggested that the N^d gene in the Romney was associated with reduced growth rate,[13] subsequent selection in the Drysdale breed appears to have overcome this disadvantage. Performance of the Drysdale at Whatawhata[19] has closely matched that of the Romney.

A 50% increase is reported[1] in lambs born per ewe lambing for Booroola Merino x local Merino 2-tooth and 4-tooth ewes over contemporary Merinos, with little difference in live weight or in fleece weight. Other studies[2][29] have demonstrated substantial increases in ovulation rates in Booroola crosses with the Merino and the Romney, relative to local straightbred controls.

Preliminary results[12] from the crossbreeding evaluation, subsequently terminated, of the "exotic" breeds imported in 1972 pinpointed the high fertility and prolificacy of Finn and East Friesian crossbred ewes and the rapid growth of Oxford and to a lesser extent German Whiteheaded Mutton crosses, relative to the Border Leicester x Romney cross. Finn-cross lambs showed higher survival rate than other crosses and grew well to achieve very satisfactory weaning and hogget body weights. The German and Oxford cross ewes produced more wool than the Romney but less than the Border Leicester x Romney, the Finn cross clipping 10% less wool than the Cheviot and Dorset crosses and 15% less than the Romney.

Caution is necessary in generalizing from the foregoing experimental findings. In some trials the adequacy of breed or sire sampling is open to question, particularly in the light of established wide genetic variation within any breed.[9] Differential genetic improvement among breeds may render past comparisons inappropriate to present stock. Further, the trials have necessarily covered only a limited range of New Zealand environments. Information is lacking on the extent to which breed performance rankings may change in different farming conditions.

It is widely believed that the Merino is prone to fleece discolouration, footrot and other health problems on wetter, more productive pastures and hence is profitable only under sparse grazing conditions in high or dry country; that the Perendale is superior to its competitors in hard hill country; that the advantages of the Coopworth can only be exploited under good grazing conditions; that the Down breeds can make no contribution to improved breeding ewe performance; but there is little objective evidence to support these beliefs.

A final but important qualification is that in most cases breeds and crosses have been compared on the basis of per animal performance; the more appropriate, but unfortunately much less measurable criterion should be production efficiency, i.e. returns per unit of feed consumed or per unit area of land. Allowance in any breed comparison is thus necessary for extra feed costs associated with greater maintenance (larger size) or production requirements.

PROSPECTS FOR THE FUTURE

The sheep originally introduced have been efficiently adapted and improved by the New Zealand breeder to serve local needs. The superior wool production and hardiness of the New Zealand Romney and meat potential of the present Southdown over their English counterparts bear testimony to the success of these endeavours, as do the established reputation of such derived breeds as the Corriedale, Perendale and Coopworth.

Breeds and breeding patterns will undoubtedly continue to change in the future in response to advances in technology, competition from alternative uses of land and other resources, changing markets and the economic need for maximum productivity.

The genetic variation existing within all breeds provides scope for selective improvement towards desired goals. It is necessary however that breeding objectives be clearly defined. The long-term success and even survival of present breeds will depend on their productive efficiency and their ability to meet the challenges of future change. Breed societies themselves will prosper only as they actively encourage the application of scientific knowledge to genetic improvement, not only within their breed but also in the national flock.

Sound choice among breeds and strains in a competititve efficiency-oriented future industry will be facilitated by the recording of performance, albeit at a simple level, in commercial flocks; by industry-supported progeny testing trials; and by the characterization of breed performance attributes from planned experiments in appropriate environments.

Much experimental and scientific evidence points to the advantages of crossbreeding over purebreeding systems, particularly in terms of female performance. More information is needed however on levels of hybrid vigour in different crosses and for different production traits. Application of crossbreeding is conditioned in large measure by operational and organisational factors which also demand investigation. Such studies will indicate the desirability and feasibility of further stratification in the sheep industry, for example the use of first cross females on improved hill country, as in the United Kingdom.

In the quest for more efficient sheep production, the potential of overseas breeds cannot be ignored.[7] For many traits, particularly those responding only slowly to selection within present breeds, introduction of favourable genes from other populations possessing clear superiority for such traits can provide a rapid means of flock improvement. This applies especially to increase in prolificacy. If the Booroola Merino lives up to its early promise it can be expected to achieve wide usage and initiate development of further new "synthetic" breeds. The genetic diversity arising from a range of breeds of varied performance attributes also confers powerful flexibility to meet changing future market requirements and production systems. Adequate animal health safeguards are of course of paramount importance in introducing new stock.

Perhaps the greatest difference between the future and the past will lie in the rationalization of breed usage and crossbreeding, based on scientific principles and factual evidence rather than tradition coupled with trial and error. In the immediate future one can predict some further expansion of the Perendale and Coopworth breeds, but at a reducing rate; their ultimate levels in the ewe population will depend in part on improvement progress already manifest in the Romney itself, in part on the usage of the Booroola Merino and perhaps of subsequent new breeds, in part on the extent to which crossbreeding practices are adopted in the industry and in part on trends in relative returns from meat and wool. Expansion of the Drysdale and other carpetwool breeds will be determined not only by export market outlets and premiums for their specialty wool, but also by their ability to compete with other ewe breeds as export lamb dams. Relative usage of different meat sire breeds will involve compromise between production systems and market requirements. The present trend from the Southdown to the larger Down breeds which sire larger, leaner lambs, may be expected to persist. However, the suitability of the Southdown in a milk-lamb production system should ensure its continued role in supplying lighter lamb carcass markets.

Acknowledgement -

The authors are indebted to the late Sir G. S. Peren and Mr T. R. Maskew (Secretary, N.Z. Sheepbreeders Association) for their valued assistance: and to the secretaries of the Romney, Southdown, Coopworth, Perendale, Drysdale and Corriedale Breed Associations.

REFERENCES

[1] Allison, A. J.; Stevenson, J. R.; Kelly, R. W., 1977: Reproductive performance and wool production of Merino and high fertility strain (Booroola) x Merino ewes. *Proc. N.Z. Soc. Anim. Prod.*, 37: 230-234.

[2] Allison, A. J. (pers. comm.)

[3] B.W.M.B. 1968: *British Sheep Breeds – Their Wool and its Uses.* British Wool Marketing Board, Isleworth, U.K. 84 pp.

[4] Carter, A. H., 1969: Research with fine wool Merinos. *N.Z. Jl Agric.,* 118(5): 22-23.

[5] Carter, A. H., 1972: New blood sought for N.Z. sheep flocks. *N.Z. Jl Agric.,* 125(1): 25-33.

[6] Carter, A. H., 1975: Importation and utilization of exotic livestock breeds. *Proc. III World Conf. Anim. Prod., Melbourne 1973.* (ed. R. L. Reid) Sydney Univ. Press, 608-615.

[7] Carter, A. H., 1976: Exploitation of exotic genotypes. In *Sheep Breeding* (Eds G. J. Tomes, D. E. Robertson and R. J. Lightfoot) Western Australian Inst. of Technology, Perth p. 117-128.

[8] Carter, A. H., Unpublished.

[9] Carter, A. H.; Kirton, A. H., 1975: Lamb production performance of 14 sire breeds mated to New Zealand Romney ewes. *Livest. Prod. Sci.,* 2: 157-166.

[10] Clarke, J. N., 1977: In *Annual Report of the Research Division* Ministry Agric. and Fisheries, Wellington p. 200-201, 228-230.

[11] Clarke, J. N.; Geenty, K. G., 1977: Export lamb production — a sire and dam breed comparison. *Aglink AST2,* Ministry of Agriculture and Fisheries, Wellington.

[12] Clarke, J. N.; Meyer, H. H., 1977: The performance of exotic sheep crosses — progress report. *Proc. Ruakura Fmrs' Conf.:* 34-41.

[13] Cockrem, F., 1963: Body growth and fleece development in the New Zealand Romney N-type sheep. *Anim. Breed. Abstr.,* 31: 445-453.

[14] Cook, J. G., 1924: Introduction of sheep into New Zealand. *N.Z. Jl Agric.* 28: 161-162.

15 Coop, I. E., 1974: Crossbreeding, interbreeding and establishing a new breed of sheep. *Proc. N.Z. Soc. Anim. Prod.*, *34:* 11-13.

16 Coop, I. E.; Clark, V. R., 1952: A comparison of breeds of ram for fat-lamb production. *N.Z. Jl Sci. Technol.*, A 34: 153-171.

17 Coop, I. E.; Clark, V. R., 1957: Breeds of rams for fat lamb production. *N.Z. Jl Sci. Technol.* A 38: 926-946.

18 Corriedale Sheep Society 1924: *Flock Book* Vol. 1.

19 Dalton, D.C., 1977: In *Annual Report of the Research Division*, Ministry Agriculture and Fisheries, Wellington, 1976-77: 79-81.

20 Dalton, D. C.; Clarke, J. N.; Rattray, P. V.; Kelly, R. W.; Joyce, J. P., 1978: Sheep breeds — comparison of Romneys, Coopworths and Perendales. *Aglink*, FPP132. Ministry of Agriculture and Fisheries, Wellington.

21 Davison, R. 1977: Export lamb survey October 1975 to December 1975. N.Z. Meat and Wool Boards' Economic Service, Paper No. T32.

22 Dry, F. W., 1955: The dominant N gene in New Zealand Romney sheep. *Aust. Jl Agric. Res.*, 6: 725-69.

23 Eastwood, K.; Marshall, P.; Wickham, G., 1976: *Sheep Breeds in New Zealand*. Ministry of Agriculture and Fisheries, Wellington, Bull. No. 414.

24 Geenty, K. G.; Clarke, J. N., 1977: A comparison of sire and dam breeds for the production of export lambs slaughtered at 3, 4½ and 6 months of age. *Proc. N.Z. Soc. Anim. Prod.*, 37: 235-242.

25 Geenty, K. G.; Jagusch, K. T., 1974: A comparison of the performance of Dorset, Corriedale and Romney sheep during lactation. *Proc. N.Z. Soc. Anim. Prod.*, 34: 14-18.

26 Hight, G. K.; Dalton, D. C., 1974: Existing sheep breeds in New Zealand and the way to exploit their potential. *Proc. Lincoln Fmrs' Conf.:* 42-53.

27 Jopp, A. J., 1969: Breeds of sheep trial conducted at Moutere, 1964-1967. Private Publn.

28 Mason, I. L., 1969: *A World Dictionary of Livestock Breeds, Types and Varieties.* (2nd edition) Commonwealth Agricultural Bureaux, U.K.

29 Meyer, H. H.; French R. L., 1979: Hogget liveweight-oestrus relationships among sheep breeds. *Proc N.Z. Soc Anim Prod.*; 39: 56-62.

30 N.S.B.A., 1968: *British Sheep.* National Sheep Breeders' Association (U.K.)

31 *New Zealand Official Yearbook,* 1893 *et seq.* Department of Statistics, Wellington.

32 N.Z.S.B.A., 1902: *Flock Book* Vol. I (New Series) New Zealand Sheep Breeders Association, Christchurch.

33 N.Z.S.B.A., 1906: *Flock Book* Vol. II (New Series) New Zealand Sheep Breeders Association Christchurch.

34 N.Z.S.B.A., 1977: *Breeds of Sheep Registered with the New Zealand Sheepbreeders Association.* New Zealand Sheepbreeders Association, Christchurch.

35 Ryder, M. L., 1964: The history of sheep breeds in Britain. *Agric. Hist. Rev.*, 12: 1-12, 65-84.

36 Sheeplan, 1977: *Sheeplan Performance Recorded Flocks.* Ministry of Agriculture and Fisheries, Wellington.

37 Stevens, P. G., 1961: *Sheep. Part 2: Sheepfarming Development and Sheep Breeds in New Zealand.* Whitcombe and Tombs Ltd., Wellington.

38 Stevenson, J. R. (pers. comm.).

39 Wickham, G. A., 1978: Development of breeds for carpet wool production in New Zealand. *World Rev. Anim. Prod.*, 14: 33-40.

3

Basic Concepts and Simple Sheep Genetics

H. T. Blair

Massey University, Palmerston North

Consultants: *A. N. Bruere, F. R. M. Cockrem, A. L. Rae*

SUMMARY

This chapter provides a basic introduction to animal genetics. The aim is to provide the reader with enough knowledge of genetics to enable him to obtain greater understanding in later chapters on selection and sheep improvement. The chapter starts with an introduction on how traits are inherited. After presenting simple cases of inheritance in sheep involving one or two allelic series the more complicated multifactorial type inheritance, which is of great importance in animal genetics, is discussed.

INTRODUCTION

To understand why selection will result in the improvement of some characteristic within a flock, a knowledge of basic genetics is required.

It has been known almost since stock became domesticated that individuals performing well, or poorly, in productive traits could pass on part of that good, or poor, performance to their progeny. During the past fifty years understanding of the method of transmission, and how this affects animal breeding has advanced rapidly.

It is now known that the dam and the sire both transmit equal amounts of information to their progeny. This transmitted information is in the form of DNA, often referred to as "genetic material". DNA is a chemical complex and is found in the nucleus of most cells, which make up the body of an animal. The amount of DNA in sperm and eggs is half of that found in other cells so that when these two fuse during fertilization the "normal" amount of DNA is again present.

At fertilization an egg and one sperm combine to form a *zygote* or fertilized egg. All of the cells which make up the foetus are derived from the original single-celled zygote. When cells divide to provide the increasing mass of the growing foetus, the DNA in the parent cell duplicates itself so that the resulting daughter cells receive an exact copy of the parent cell's DNA. Because of this method of division all cells making up the body must have exactly the same DNA and therefore all cells in an individual have the same information from which their functions are determined.

During the development and growth of the animal only parts of the available information contained in the DNA are "read", hence cells of different shape, function and use are formed. How the reading occurs is complex and will not be discussed in this chapter.

The final appearance and performance of any individual is determined by an interaction between the genetically supplied information and the environment. This interaction is a continuous process from the time of fertilization until the death of the animal. Environmental effects include factors such as; dam's nutrition during pregnancy, milk production of the dam, any disease or parasites that the animal is subjected to, climate and stock management. The effects of the environment have been shown in experiments involving monozygous twins which, because of the manner in which they arise, have identical sets of genetic information (DNA). By subjecting each of the twins to different environmental effects, such as varying nutritional levels, it has been shown that different levels of performance can be achieved.

GENES AND CHROMOSOMES

The DNA within cells making up the body of an animal exists in an ordered form. During cell division it is combined with other chemical compounds to form many thread-like strands called *chromosomes*. When the DNA duplicates itself the chromosomes also duplicate. Therefore after division both daughter cells have the same number of chromosomes that the parent cell originally had before division. Fig. 3.1 depicts the process in a cell with four chromosomes.

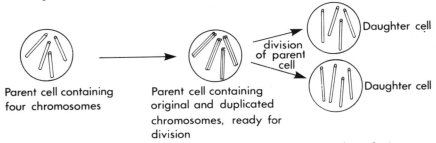

Parent cell containing four chromosomes

Parent cell containing original and duplicated chromosomes, ready for division

division of parent cell

Daughter cell

Daughter cell

Fig. 3.1: Diagrammatic representation of the chromosome numbers during duplication.

The only cells in the body that do not have a full complement of chromosomes are the sperm or egg cells. When the DNA content is halved the chromosome number is also halved; in the above example if the parent cell had been dividing to form sperm or ova the resulting cells after division would only have had two chromosomes. The number of chromosomes is relatively constant between different individuals of the same species, e.g. sheep normally have 54, cattle have 60 and humans have 46. However, some individuals have

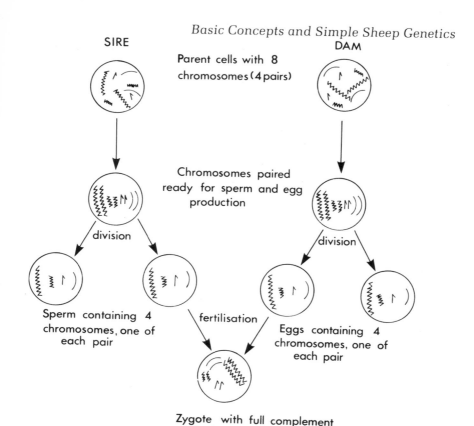

Fig. 3.2: Diagrammatic representation of chromosome distribution during production of sexual cells and fertilization.

abnormal numbers of chromosomes and sheep with 53, 52 or 51 chromosomes have been bred.[10] The reduced number is probably caused by the union of some chromosomes.

A feature of chromosomes is that they normally occur in pairs. Thus sheep have 27 pairs and cattle have 30 pairs of chromosomes. When pairing of chromosomes occurs during the life of any cell usually two that closely resemble each other pair off. During sperm and ova formation members of a pair separate so that in any one sperm or egg only one member of each pair is represented. Thus when an egg and a sperm fuse at fertilization the chromosomes will again be able to pair off with a like kind.

After pairing occurs in male mammalian cells it is noticeable that the members of one chromosome pair are quite dissimilar. This pair is responsible for sex determination. The larger chromosome is called the X chromosome and the smaller one the Y chromosome.

Female mammals carry two X chromosomes hence eggs have X chromosomes only. Sperm carry either X or Y chromosomes. The sex of

41

any offspring is therefore determined by the type of sperm fertilizing the egg.

Because sperm carrying X or Y chromosomes are produced in equal numbers, and fertilization is a random event, the numbers of males and females that are conceived are approximately equal.

Fig. 3.2 shows the reduction of chromosome number in the production of ova and sperm, followed by fertilization, at which the full complement of chromosomes is regained.

From the preceding discussion it should be clear that there are two types of cell division. *Mitotic divisions*, where a parent cell divides to give two daughter cells with the same number of chromosomes as the parent, are responsible for increasing the number of cells in the body of an animal as it grows. *Meiotic divisions* occur almost exclusively in the testes or ovaries and are responsible for the production of cells with half the normal body complement of chromosomes.

Each chromosome consists of a great number of *loci*, each of which contains a portion of DNA carrying information for the building of a particular part of the body. The two members of any particular chromosome pair can be divided into exactly the same loci, see Fig. 3.3.

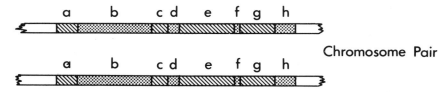

Fig. 3.3: Diagram illustrating the distribution of corresponding loci along parts of a chromosome pair.

It is a combination of the information carried at any two corresponding loci that determines some basic structure within the body, e.g. the a's may combine to determine whether or not an individual has horns and the two f's may combine to determine wool colour, etc. Each of these loci is occupied by a gene. Thus a *gene* is a part of a chromosome, which in conjunction with its partner on the other chromosome pair, determines some basic structure in the animal.

The information contained by any two corresponding genes need not be the same. If the information is the same at the two corresponding genes the individual is said to be in a *homozygous* state for that particular pair of genes. If the information is different then the individual is in a *heterozygous* state for that gene pair. The question then arises as to how the information is interpreted. If an animal is homozygous for a particular gene pair then there is no problem as the two pieces of information agree and interpretation is straightforward. But if the pair of genes are in a heterozygous state and if the information from one gene is used in preference to that on the corresponding gene, the gene from which the utilized information is supplied is said to be *dominant* over the other gene. The gene that carries the unused information is deemed *recessive*. This order of dominance and recessiveness, once established, normally does not change within the individual or within the population. The case where one gene has no influence at all is not common in animal genetics. Usually the information on both genes will be used, although one gene may have more influence than the other. In this case that gene is said to be *incompletely* or *partially dominant*.

The genes that occur at corresponding points on any given pair of chromosomes are referred to as *alleles*. Thus in Fig. 3.3 the two genes at the b loci are alleles, as are the two at the c loci. But gene 'c' on one chromosome and gene 'f' on the other are not alleles, as they occur at different positions on the chromosome pair. Within any one animal only two alleles can possibly occur, because there is only one position on each of a pair of chromosomes. Within the species there can be several alleles that might occupy these two positions; which alleles fill the positions is determined by the alleles the individual's dam and sire carried and by chance allocation of one allele from each parent.

Because any one chromosome tends to be passed as a whole between generations, genes occuring on that chromosome will be transmitted together; such genes are said to be *linked*. During the production of sperm or egg cells it is possible for linked genes to become separated due to the breaking and reunion of chromosomes. The members of a chromosome pair break in the same position and exchange equivalent pieces of genetic material. Fig. 3.4 illustrates this process.

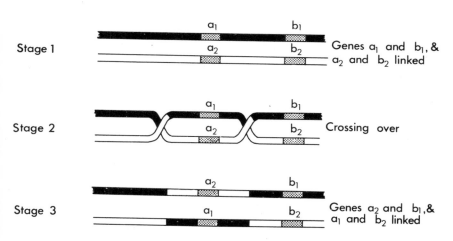

Fig. 3.4: Diagram illustrating crossing over.

Crossing over is a frequent occurrence in sperm and egg production, thus genes that are not close together on the chromosome have a good chance of being separated by crossing over. Conversely genes that are close together will have little chance of being separated by crossing over.

An example of linkage is found in Russian Precoce Merinos and in Australian Merinos where the genes for polledness and *cryptorchidism* (incompletely descended testicles) are linked.[5] Poll rams in these breeds have a high incidence of cryptorchidism.

A SIMPLE PAIR OF GENES, ONE RECESSIVE AND ONE DOMINANT

As an introduction to the inheritance of characters in a flock of sheep a simple case of one pair of alleles will be considered, i.e. for this example there will only be two possible sets of information that can occur at the locus being discussed. To simplify things further one allele will be completely dominant over the other allele, so that when the heterozygote is encountered only the dominant set of information is interpreted and expressed.

The Recessive (nr) Gene for Medullation

Between 1930 and 1960 Dr F. W. Dry worked with a selected group of Romneys in an attempt to determine the mode of inheritance of hairy fibres. One end result of his work is the well-known, hairy-fleeced Drysdale breed. He attributed the hairy fleece characteristic to a series of genes, of which the partially dominant N gene (now known as N^d) is the most important.[18] Not so well known as the N^d gene is the nr gene,[19] which occurs at a different locus, i.e. they are not alleles. The nr gene also carries information for a hairy fleece, but it is recessive to its other allele the $+^{nr}$ (referred to as the + gene in the following discussion) gene, which codes for the "normal" Romney type fleece with few hairy fibres.

In a flock that contains the nr gene three possible arrangements of the 2 alleles can occur on the chromosome pair. Rather than drawing the actual chromosomes each time the alleles are referred to, an abbreviation is to note the alleles being discussed:

| (a) | (b) | (c) |
| + + | + nr | nr nr |

Individuals carrying two + genes will have non-hairy fleeces. Those having one + gene and one nr gene will also have non-hairy fleeces. Since they carry an nr gene, but do not show any expression of hairy fibres, they are often termed *carriers*. Individuals carrying set (a) or set (b) will be indistinguishable by sight but may be separated by controlled breeding experiments; (c) represents hairy fleeced sheep as the recessive gene can now be expressed in the absence of a + gene.

The mode of transmission between generations can be represented in tabular form. As an example consider a homozygous dominant sire mated to a homozygous recessive ewe:

TABLE 3.1: Types of Progeny Resulting from Mating a + + Non-hairy Ram to a nr nr Hairy Ewe

| Ova types | Sperm types | |
	+ (a)	+ (b)
nr (c)	+ nr	+ nr
nr (d)	+ nr	+ nr

All progeny are heterozygous (carriers) and therefore non-hairy.

The sire has two + genes, each on a member of a chromosome pair. In the formation of sperm the pair part company, each going to a different sperm. Sperm (a) and (b) each have an equal change of fertilizing egg (c) or egg (d), conversely egg (c) has an equal chance of being fertilized by sperm (a) or by sperm (b). Therefore each of the four possible zygotes has an equal chance of being produced. Table 3.1 shows that all progeny will be heterozygous (carriers) and therefore non-hairy.

Another type of mating occurring in a flock with the nr gene is a heterozygous ram over a homozygous recessive ewe.

TABLE 3.2: Types of Progeny Resulting from Mating a + nr Non-hairy Ram to a nr nr Hairy Ewe

| Ova types | Sperm types | |
	+	nr
nr	+ nr	nr nr
nr	+ nr	nr nr

As can be seen from Table 3.2, half of the progeny from such a mating will be heterozygous (non-hairy) and half will be homozygous recessive (hairy).

If the sexes were swapped in either of the two examples shown, the resulting progeny would remain unchanged.

The matings shown in Tables 3.1 and 3.2 supply a method of differentiating between homozygous dominant and heterozygous individuals, both of whom have non-hairy fleeces. The procedure is to mate the non-hairy-fleeced animal to a number of known nr nr animals. If no hairy fleeces occur in the progeny then the animal being tested should be homozygous dominant, i.e. + + (see Table 3.1). If one or more hairy fleeces occur then the animal must be heterozygous, i.e. + nr (see Table 3.2).

It should be emphasised here that these results and explanations only hold true for the recessive nr gene and that the dominant genes for hairiness give rise to entirely different results.

A third type of mating that could occur with the same two allelic genes is a heterozygous ram crossed with a heterozygous ewe.

TABLE 3.3: Types of Progeny Resulting from Mating a + nr Non-hairy Ram to a + nr Non-hairy Ewe

Ova types	Sperm types	
	+	nr
+	+ +	+ nr
nr	+ nr	nr nr

From Table 3.3 it can be seen that it is possible to obtain hairy-fleeced progeny when both of the parents are carriers. Of the four possible combinations in the progeny two will be heterozygous, one will be homozygous dominant, and one will be homozygous recessive. Therefore from this type of mating the expected ratio of progeny is 3 non-hairy : 1 hairy.

As well as the three discussed matings there are three other types of matings which could occur but will not be discussed in depth as their outcomes are easy to follow.

(i) + + x + +. The only possible outcome from this mating is non-hairy fleeced progeny because both chromosomes in the dam and the sire carry dominant genes.

(ii) nr nr x nr nr. All progeny from this mating will have hairy fleeces.

(iii) + + x + nr. All progeny will be non-hairy, the single recessive gene will have its effect covered by a dominant gene. Half of the progeny will be heterozygous.

Congenital Photosensitivity in Southdowns

This syndrome has been reported in New Zealand[23] and Californian[13] [27] Southdowns.

Lambs with congenital photosensitivity appear normal at birth, the first visible signs show at about the time the lamb begins to eat grass. The disease is very similar in its effects to facial eczema, except that there is no spontaneous recovery following the period of photosensitivity. As the disease progresses the ears are often lost and the eyes may be destroyed; lambs thus affected will die within two to three weeks after the onset of the disease. The cause of death is starvation through the inability to feed. Animals permanently sheltered from the sun grow at rates comparable to their unafflicted contemporaries; under shelter, affected animals also breed and reproduce normally.

The primary cause of congenital photosensitivity was described by Clare;[12] those interested should refer to his article. Breeding experiments showed that congenital photosensitivity was inherited as a simple recessive gene.[21] The condition is an example of a *sub-lethal* trait because the animal dies only under certain conditions (for animals to die of congenital photosensitivity they must be exposed to sunlight). If the foetus dies during pregnancy, or if the animal is to die under any conditions then the trait is considered to be lethal.

If a heterozygous ram is mated with a heterozygous ewe the pattern of inheritance will be similar to that shown in Table 3.3. Three out of every four progeny born by the mating of two carriers will be normal, but two out of the three normal lambs will be carriers of the sub-lethal gene. Because of the undesirable nature of this condition breeders should, when possible, cull carriers as well as affected individuals. The problem is how to identify carriers.

Recognizing Carriers of Undesirable Recessives

Identifying carrier rams will reduce the incidence of the disease faster than will identifying carrier ewes, since rams leave more progeny than do ewes. One method of identifying these unwanted rams is to progeny test them as ram hoggets.

Progeny tests for recessives are most accurate when the ram is mated to ewes that are homozygous for the recessive factor. However, these may be difficult, if not impossible, to maintain. Alternatively, if there is a number of known carrier ewes, the ram could be mated to these. Another method is to mate the ram to half sisters. This is a test for all types of recessives. While this latter method prevents recessives being passed from one generation of sires to the next, the number of maternal half sisters is insufficient to allow accurate tests that will show whether a sire has inherited an undesirable recessive from his dam. If any affected progeny are produced the ram must be a carrier and should be culled, together with his half sisters if this is possible.

Animals carrying recessive genes for some traits can be identified by biochemical tests. Although the heterozygous individuals do not appear different from normal animals, there may be some upset in their biochemical pathways. If this upset can be traced a biochemical test may be devised for use in identifying carrier animals. Although this type of testing is not yet used with sheep it is being used in cattle to identify pseudolipidosis carriers.

Dwarfism

Two other examples of simply inherited traits are dwarfism and the Ancon Dwarf. Dwarfism was first reported about 1940 in Missouri.[6][7] It occurred in the Southdown breed and appeared to be inherited as a simple, recessive, *lethal* gene. The lambs were born with short legs, bulging forehead and were thick through the shoulders. All lambs thus afflicted died within four weeks of birth. Dwarfism also appeared in 1965 in New South Wales.[11] The Dorset Downs which were affected had an overall reduction in size giving them a thick-set appearance.

The Dorset Down dwarfs survive to maturity thus the gene involved may be different from that described in Southdowns.

The *Ancon Dwarf* is an animal with a normal-sized body but with short legs. It was first thought to have occurred about 300 years ago in Massachusetts. Initially it was considered a useful trait because the short-legged animals could not jump over the low stone walls then used for subdivision. Unfortunately the joints of these animals were prone to crippling, thus limiting the use of this line of sheep. They finally died after about a century of use. The Ancon Dwarf has since re-appeared in a Cheviot flock in Norway,[22] and in a Merino flock in Texas.[32]

OVERDOMINANCE

In examples previously discussed one gene has always been completely dominant over its allelic partner. In some cases there may be mixing of the information supplied by the two different genes. If this occurs it will be possible to distinguish between the heterozygote and the homozygotes. Occasionally the heterozygote offers some kind of advantage over either of the homozygotes; this is referred to as *overdominance*.

The Wensleydale breed in England supplies a simple example of overdominance. The problem facing breeders of these sheep in the 1920's was that approximately one lamb in five was black, even though all ewes and rams had white fleeces.

The interpretation of breeding records and experiments[16] established that the expression of black fleeces required the presence of two recessive genes, while white fleeces resulted from the presence of one or two 'dominant' genes. But the gene for white wool was not completely dominant. In the heterozygote certain areas of the skin, such as the ears and the face, had a dark tinge (called blue) because of the partial expression of the 'recessive' gene.

One of the standards set by the Wensleydale Longwool Sheep Breeders' Association was that ewes and rams used for breeding should have a deep blue colour in the face and ears. By selecting such animals Wensleydale breeders were inadvertantly maintaining flocks of heterozygous rams and ewes. Through the mating of heterozygotes one out of every four progeny should be homozygous for the recessive gene (see Table 3.3) and have black fleeces; the figure of one-fifth quoted previously, is close enough to verify this argument considering that some homozygous whites would be retained to maintain flock numbers.

MULTIPLE ALLELES

The number of alleles that can occur at any particular locus is not limited to two. Several genes can occur at a locus, but only two of these alleles can occur in any one individual.

Horn Growth Genes

It appears that there are three alleles controlling the growth and development of horns in Merinos.[15] Of these three alleles P codes for the poll condition, p' for horns and p for sex-limited horns.

Table 3.4 shows the effects of particular gene combinations according to sex.

TABLE 3.4: Resulting Expressions from Various Combinations of Poll and Horn Genes

Genes Present	Expression			Breeds
	Rams	Wethers[1]	Ewes	
PP	Poll	Poll	Poll	British Downs and Longwools, Poll Dorset and Poll Merino
Pp'	Small Horn[2]	Small Horn	Small Horn	
Pp	Small Horn	Poll	Poll	
p'p'	Horns	Horns	Horns	Dorset Horn
p'p	Horns	Horns	Horns	Some Merinos
pp	Horns	Poll	Poll	Most Merinos

[1] Male lambs castrated soon after birth.
[2] Small horns are horns with a mature length of less than 8 cm.

While pp rams have horns, pp wethers and ewes do not. This difference in expression is due to the presence, or absence, of male sex hormones. These hormones in pp rams can promote the growth of horns, or in Pp rams the growth of small horns. In wethers the removal of the testes inhibits the production of male sex hormones; pp and Pp wethers do not develop horns after castration.

The poll gene P is only partially dominant over the horn gene p' as can be seen from the presence of small horns in Pp' ewes and wethers. But p' is completely dominant over the other poll gene p, since heterozygous ewes (pp') produce horns.

The three horn and poll genes have in the past been designated different letters. Table 3.5 contains three different naming systems. Dolling's second[15] nomenclature is now generally accepted.

TABLE 3.5: Nomenclature of the Horn and Poll Genes

Dominant Poll Gene	Horn Gene	Recessive Horn Gene	Author
H	H'	h	Warwick et al.[33]
P	P'	p	Dolling[14]
P	p'	p	Dolling[15]

Development of Poll Strains from Horned Breeds

Poll strains of horned breeds can be developed by introducing the P gene from a poll breed, then back-crossing to the horned breed and culling progeny with full-sized horns. An example of this was the formation of the Poll Dorset strain from the Dorset Horn. Firstly Dorset Horn ewes were mated with Ryeland or Corriedale rams, since Ryeland and Corriedale sheep are homozygous for the dominant poll gene P.

TABLE 3.6: Horn Gene Combinations in First Cross Progeny of Dorset Horn Ewes (p'p') × Corriedale or Ryeland Rams (PP)

Ova type	Sperm type	
	P	P
p'	Pp'	Pp'
p'	Pp'	Pp'

All progeny were heterozygous and therefore had lumps or small horns and were half Dorset Horn and half Ryeland, or Corriedale.

The first cross rams were then mated back to Dorset Horn ewes or Dorset Horn rams with first cross ewes.

TABLE 3.7: Horn Gene Combinations when Mating Dorset Horn Ewes (p'p') × First Cross Rams (Pp')

Ova type	Sperm type	
	P	p'
p'	Pp'	p'p'
P'	Pp'	p'p'

Half of the progeny from this cross were horned p'p' and were culled. The progeny from this cross were three-quarters Dorset Horn and one-quarter Ryeland or Corriedale. The next back cross involved mating the heterozygous three-quarter Dorset Horn sheep back to Dorset Horns to further increase the proportion of Dorset genes in the line. This utilized the same procedure as is shown in Table 3.7. Once again, half the progeny were horned (culled) and half were heterozygous Pp'. All progeny from this cross were seven-eighths Dorset and one-eighth Ryeland, or Corriedale. The heterozygous animals were then interbred to stabilise the Poll Dorset strain.

TABLE 3.8: Horn Gene Combinations when Interbreeding Heterozygous Seven-Eighths Dorset Ewes × Seven-Eighths Dorset Rams

Ova type	Sperm type	
	P	p'
P	PP	Pp'
p'	Pp'	p'p'

One-quarter of the progeny were pure-breeding Poll Dorsets (PP). One-half of the progeny were heterozygous and could be interbred again or mated to PP to produce more PP individuals. One-quarter of the progeny would be horned p'p' and hence culled.

In the normal Merino most sheep are homozygous pp but about 10% are p'p or p'p'. The Poll Merino has developed through selection for the polled gene which apparently arose by mutation in a horned Merino flock.

The horns in carpet wool sheep are not thought to be determined by the "P series". They appear to arise because of a multiple effect of the genes at the N locus or of another gene closely linked to N.

Carpet Wool Genes

The inheritance of hairy fibres in sheep is determined by a number of different genes. The role of the recessive nr gene has already been discussed. Apart from this gene there is another series of genes, non-allelic to the nr gene,[18] that are involved in determining the hairiness of the fleece in Romneys and related breeds. The four alleles identified so far in this series are the N^t, N^j, N^d and n genes.

The N^d and n genes were first reported by Dry in 1955.[17] He referred to them as the N and + genes. He suggested that the N^d gene, which codes for a high density of hairy fibres in the fleece, was incompletely dominant over the n gene, which codes for a normal Romney fleece with few hairy fibres. At certain positions on the body the N^d gene is almost completely dominant, e.g. at the mid-dorsal line on the level of the last rib. Behind the shoulder there is a small patch at which the N^d gene is recessive, as a dense hairy coverage does not occur in this position unless the animal is homozygous N^dN^d. The standard back position and the shoulder patch are used in differentiating between N^d N^d, N^dn and nn lambs. If the lambs have a dense cover of hairy fibres in the standard back position then they must be N^dn or N^dN^d; if not then they must be nn. N^dn and N^dN^d lambs are distinguished by observing the shoulder patch; if they have a dense coverage of hairy fibres then they are N^dN^d; if not they are N^dn. Using these methods Dry established a flock of pure-breeding, hairy fleeced sheep which became the basis of the Drysdale breed.

Since the establishment of the Drysdale two other hairy-fleeced

51

breeds have been founded. These are the Tukidale and Carpetmaster (See photo 3.1).

Photo 3.1: From left, homozygous $N^d N^d$ ewe, heterozygous N^j n ewe, heterozygous N^t n ewe.

The carpet-type fleece of these breeds has been shown to be due to genes allelic to the N^d gene of the Drysdale.[34]

It is not certain whether the genes of the Tukidale and Carpetmaster are the same or different. There is no common ancestry between the two breeds and the same rare mutation must have occurred in two instances if the same gene is shared.

Thus there is an allelic series of three or four genes determining fleece type and these have been named[34]—

N^t	Tukidale	}completely dominant and indistinguishable
N^j	Carpetmaster	
N^d	Drysdale	partially dominant to n
n	Romney	recessive.

Since the N^t and N^j genes are completely dominant it is not possible to tell the difference between homozygous dominant and heterozygous animals, i.e., between $N^t N^t$ and $N^t n$, and $N^j N^j$ and $N^j n$. This will lead to problems when establishing a purebreeding flock of hairy-fleeced sheep. It will be very difficult to eliminate the last few n genes from the flock. While the n gene is present there will always be the chance of some Romney-type fleeces turning up.

The fact that the N^t, N^j and N^d genes induce horn growth as well as hairy fleeces is an example of pleiotropy, i.e. multiple effects of the one gene.

MUTATIONS

A *mutation* is a change in the genetic material that results in any gene changing its expression. Mutations arise from either physical rearrangements of the genetic material or from actual chemical changes within the gene. These changes are occurring continually in the genetic material but they are not often expressed for two reasons:

(1) any mutations causing major changes in gene expression are unlikely to result in a viable zygote at fertilization; i.e., major mutations are usually lethal at an early stage.

(2) mutations are often recessive and therefore will be hidden by a dominant allele from the other parent.

Mutations are rare in gametes or cells that produce the gametes. It is unusual to get exactly the same mutation process at the same locus in different sheep. Therefore if the mutant gene is recessive it is only by the mating of related animals, who are descendants of a parent with the mutant gene, that the mutation might become homozygous and thus be expressed. This is one of the reasons why inbred lines have a lower than normal viability; i.e., with inbreeding there is a greater chance of harmful recessive mutants becoming homozygous, thus reducing the viability of the zygote, foetus, and newborn lamb.

INTERACTION BETWEEN NON-ALLELIC GENES

In previous examples presented in this chapter only the action of genes filling corresponding loci on chromosome pairs have been considered in causing any final expression. These simple cases of one gene pair giving one final characteristic are unusual in animal genetics. Most characteristics in domestic animals result from the interaction of many different gene pairs that are at different loci and usually on different chromosome pairs. If this interaction between non-alleles results in any major modification in the expression of the gene pair then the interaction is referred to as *epistasis*.

An example of epistasis is supplied by the inheritance of fleece colour and fleece colour patterns in sheep. This has been reviewed in several publications.[1][2][4][8][9][28][30][31]

Inheritance of Fleece Colour

It is now widely accepted that fleece colouring is controlled by two main allelic series. The "A series" controls the pattern of colouring in the fleece. Adalsteinsson has recognised 6 alleles at this locus in Icelandic sheep. The "B series" determines whether any pigmentation that is present will be black or brown (Moorit).

Tables 3.9 and 3.10 show the major genes identified in each series.

TABLE 3.9: The "A Series" of Genes for Fleece Pattern

Allele	Pattern	Dominant to	Recessive to
A^{wh}	Solid white	All	None
A^{gw}	Grey mouflon	A^g, A^w, a	A^{wh}
A^g	Grey	a	A^{wh}, A^{gw}
A^b	Badgerface	a	A^{wh}
A^w	Mouflon	a	A^{wh}, A^{gw}
a	Solid black or brown	None	All

TABLE 3.10: The "B Series" of Genes for Fleece Colour

Allele	Colour	Gene interactions
B	Black	Completely dominant over b
b	Brown (moorit)	Completely recessive to B

The dominant A^{wh} gene inhibits black or brown pigment production over the whole body; therefore sheep carrying one or two A^{wh} genes will usually have white fleeces. Because of the inhibition of pigment production by the A^{wh} gene neither of the "B series" genes can be expressed.

The four intermediate alleles, A^{gw}, A^g, A^b and A^w, all cause the inhibition of pigment production in certain areas of the body, thus resulting in characteristic patterns that can be associated with each gene. The colour produced in the pigmented areas will depend on which of the "B series" genes are present. The dominance relationships within these four A series alleles are such that inhibition of pigment production dominates over pigment production. When the mouflon gene, A^w (sometimes called the reversed-badgerface gene) is homozygous, or combined with the recessive a gene, the mouflon pattern is produced as seen in photo 3.2.

The A^w gene inhibits pigment production from the belly back up to the underside of the tail and forwards from the belly up the chest to the underside of the jaw. The rest of the fleece is black or brown depending on which of the "B series" genes are present.

The badgerface colour distribution is virtually a reversal of this with the areas that are white in the mouflon being pigmented in the badgerface and vice-versa. The area covered by pigmented fibres can be variable in both the mouflon and badgerface patterns.

It is usually very difficult to distinguish between sheep which are genetically grey (A^gA^g or A^ga) and those which have lost pigmentation from environmental influences. In sheep having coarse outercoat and fine undercoat fibres the grey sheep generally have a pigmented outercoat and white undercoat,[1] the mixture giving the grey effect. Distinct outer- and inner-coats are not present in most New Zealand

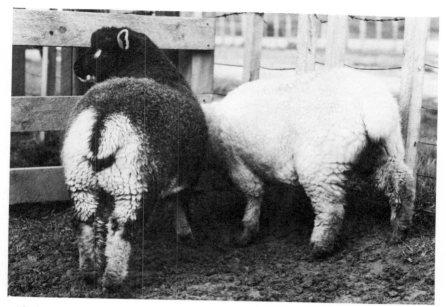

Photo 3.2: A mouflon-pattern lamb and a white lamb. The white lamb is presumably A^{wh}-BB with the - being any other A allele. The mouflon-pattern lamb will be A^wA^wBB or A^waBB.

sheep. Grey sheep may range from almost completely white to almost completely pigmented.

The recessive gene, a, causes no inhibition of pigmentation. Consequently animals homozygous for the a gene have solid colouring of either black or brown, depending on which B or b genes are present. The brown gene, b, is probably not very common in New Zealand. Many coloured sheep appear to be brown, but this is mostly a bleached black colour. A true idea of the fleece colour is best obtained at birth before the weather has had a chance to act on the wool.

In summary the following "A series" gene combinations may occur in New Zealand.

1. White sheep must be one of the following: $A^{wh}A^{wh}$; $A^{wh}A^g$; $A^{wh}A^w$; $A^{wh}A^b$; $A^{wh}a$.
2. Mouflon sheep are either A^wA^w or A^wa.
3. Badgerface sheep are either A^bA^b or A^ba.
4. Grey sheep may be either A^gA^g or A^ga.
5. Black or brown sheep must be aa.
6. The combinations A^wA^b, A^wA^g and A^bA^g may also occur. The pattern of these can generally be predicted considering the fact that inhibition of pigment production is dominant.

The "B series" then interacts with these above combinations to determine the colour of any pigmentation, when the A^{wh} gene is absent.

With each gene pair presented above one of the following "*B* series" combinations must occur, either *BB* (black), *Bb* (black) or *bb* (brown). Therefore, as an example, a black mouflon may have any of the following combinations: A^wA^wBB (true breeding black mouflon); A^wA^wBb; A^waBB or A^waBb.

The true inheritance of fleece colour has been simplified in this example as there are at least three other gene series interacting with the two already described. The other three series are called the *C*. *E* and *S* series; they will not be discussed in this chapter as their influence is not known in New Zealand. In this discussion on fleece colour inheritance no mention is made of the inheritance of coloured limbs and head, as in the Down breeds, or inheritance of black spots which often appear on the body of otherwise white sheep.

INTERACTION BETWEEN GENES AND THE ENVIRONMENT

Not all characteristics expressed in an animal are due solely to the effect of the genetic material. Often the environment has a large role to play in determining how a characteristic is finally expressed. In the previous section of this chapter the genetic control of colour was discussed. It has been shown that in some circumstances the fleece colouring can be reversed from white to black, or from black to white, in a particular animal. Since the change occurs in one animal the genetic material must be the same and therefore the change in colour must be due to an environmental effect.

Black-fleeced sheep can be induced to lose pigmentation by feeding them on a copper-deficient diet.[3] Copper is the active centre of an enzyme, tyrosinase, which is involved in the production of *melanin*, the substance responsible for pigmentation.[29] This same effect on black sheep can be obtained by feeding them a zinc-deficient diet, as zinc is also involved in melanin formation. Therefore even though an animal may be homozygous for the recessive pattern gene it is possible to obtain a fleece with little pigmentation through environmental effects. It is possible to restore the black fleece colour by supplying a diet adequate in copper, or zinc. With this information in mind black sheep can be used as on-farm indicators of copper or zinc deficiencies. Interpretation of a whitening fleece on black sheep must be undertaken carefully, since whitening can also result from bleaching by the weather.

Further examples of environmental effects on fleece colouring are supplied from experiments carried out on Suffolks and black Merinos.[25] Temporarily freezing an area of skin reversed the colour of some of the fibres growing in that region.

Wool pigmentation was changed from white to black in Hampshires by feeding them vitamin A and D deficient diets.[24] The wool colour reverted to white again when vitamins A and D were added to the diet.

Other environmental effects occur continually. For example it is well known that sheep on a high plane of nutrition will be heavier and will grow more wool than if they were on a low plane of nutrition.

It is these environmental effects that make an accurate prediction of the genetic makeup (genotype) of any individual difficult. The only measure of the genotype often available is a measure of the level of performance or appearance (phenotype) in a particular trait. If the phenotype is greatly affected by the environment, it will not supply an accurate measurement of the genotype unless some measure of the environmental effect is available.

MULTIFACTORIAL INHERITANCE

Multifactorial inheritance is the inheritance of any character dependent on several genes from different loci for its final expression.

The matings in Table 3.11 were carried out by Dry[17] and show the nature of halo-hair inheritance in Romneys without N or nr genes. The density of halo-hairs on the back of newly-born non-N and non-nr lambs was graded on a I to VI scale, grade I having no halo-hairs and grade VI having many halo hairs.

TABLE 3.11: Inheritance of Halo-Hair Abundance on the Back of New-born Lambs: The Number of Lambs with Various Halo Hair Grades According to Parent Grade

Parent Grades	Lamb Halo Hair Grades					
	I	II	III	IV	V	VI
V x V	—	3	12	10	12	5
IV x IV	—	3	2	—	—	—
IV x III	—	4	7	3	2	—
III x II	3	16	2	—	3	1
II x II	9	38	10	3	—	—
I x I	52	23	2	—	—	—

Because of the variation shown in the pattern, Dry suggested that several different gene series, all exerting some small influence, controlled halo-hair inheritance in these sheep. Thus the inheritance of hairy birthcoat in lambs is controlled by the major gene series at the N locus and also by the nr locus but is further influenced by a number of genes whose individual effects are relatively small.

With multifactorial inheritance there are no clearly-defined boundaries in physical expression between animals of slightly different genetic make-up. Rather than expressing just a few clearly defined values, as happens with characters under the control of a few main genes, characters inherited in a multifactorial fashion show continuous variation.

Most economically important traits appear to be inherited multifactorially. For example, body weight and fleece weight assume a great variety of values within one flock and hence appear to be influenced by many genes.

Because these traits are under the control of many genes it is not possible to determine which genes are present in any individual. This is in contrast to the case of colour inheritance where a black-fleeced sheep is known to have the genes *aaBB* or *aaBb*.

Thus, although simple Mendelian genetics provides the basis of animal breeding, in practice it is usually not possible to distinguish the effects of individual genes. For this reason animal breeders have developed methods which, in general, do not require knowledge of the effects of specific genes. It is these methods that are discussed in the next few chapters. However, the standard methods of animal breeding are inefficient in situations where main genes can be recognised.

REFERENCES

[1] Adalsteinsson, S. 1970: Colour inheritance in Icelandic sheep and relation between colour, fertility and fertilization *J. Agric, Res. Icel.* 2(1): 3-135.

[2] Adalsteinsson, S. 1974: Colour inheritance in farm animals and its application in selection. *Proc. 1st World Congr. Genetics* 29-37.

[3] Ashton, W. M. 1970: Trace elements in enzyme systems with special reference to deficiencies of copper and cobalt in some animal diseases. *Outlook on Agric.* 6: 95-101.

[4] Berge, S. 1974: Sheep colour genetics. *Z. Tierzuchtg. Zuchtgsbiol.* 90: 297-321.

[5] Bishop, M. W. H. 1972: Genetically determined abnormalities of the reproductive system. *J. Reprod. Fert., Suppl.* 15: 51-78.

[6] Bogart, R.; Dyer, A. J. 1942: The inheritance of dwarfism in sheep. *J. Anim. Sci.* 1: 87.

[7] Bogart, R. 1946: Inheritance and physiology of dwarfism in sheep. *Genetics* 31: 211-212.

[8] Brooker, M. G.; Dolling, C. H. S. 1965: Pigmentation of sheep. I. Inheritance of pigmented wool in the Merino. *Aust. J. agric. Res.* 16: 219-228.

[9] Brooker, M. G.; Dolling, C. H. S. 1969: Pigmentation of sheep. II. The inheritance of colour patterns in black Merinos. *Aust. J. agric. Res.* 20: 387-394.

[10] Bruere, A. N. 1975: Further evidence of normal fertility and the formation of balanced gametes in sheep with one or more different Robertsonian translocations. *J. Reprod. Fert.* 45: 323-331.

[11] Chorlton, S. 1966: Dwarfism in sheep. *Wool Technology and Sheep Breeding* 13(1): 83-84.

[12] Clare, N. T. 1945: Photosensitivity diseases in New Zealand. IV. *N.Z. Jl Sci. Tech.* A27: 23-31.

[13] Cornelius, C. E.; Gronwall, R. R. 1968: Congenital photosensitivity and hyperbilirubinemia in Southdown sheep in the United States. *Amer. J. Vet. Res.*, 29: 291-297.

[14] Dolling, C. H. S. 1961: Hornedness and polledness in sheep. IV. Triple alleles affecting horn growth in the Merino. *Aust. J. agric. Res. 12:* 351-361.

[15] Dolling, C. H. S. 1970: *Breeding Merinos.* Rigby Ltd, Adelaide.

[16] Dry, F. W. 1936: The genetics of the Wensleydale breed of sheep. II. Colour, fertility and intensity of selection. *J. Genetics 33:* 123-134.

[17] Dry, F. W. 1955: Multifactorial inheritance of halo-hair abundance in New Zealand Romney sheep. *Aust. J. agric. Res, 6:* 608-623.

[18] Dry, F. W. 1955: The dominant N gene in New Zealand Romney sheep. *Aust. J. agric. Res. 6(5):* 725-769.

[19] Dry, F. W. 1955: The recessive N gene in New Zealand Romney sheep. *Aust. J. agric. Res. 6(6):* 833-862.

[20] Dry, F. W. 1958: Further breeding experiments with New Zealand Romney N-type sheep to 1956. *Aust. J. agric. Res, 9:* 348-362.

[21] Hancock, J. 1950: Congenital photosensitivity in Southdown sheep. *N.Z. Jl Sci. Tech. A32:* 16-24.

[22] Landauer, W.; Chang, T. K. 1949: The Ancon or Otter sheep. History and Genetics. *J. Hered. 40:* 105-112.

[23] Leslie, A. 1936: A note on an obscure skin affection of Southdown lambs. *Cyclostyled report to Lincoln Agricultural College.*

[24] Light, M. R.; Klosterman, E. W.; Buchanan, M. L.; Bolin, D. W. 1952: The effect of nutrition upon wool pigmentation. *J. Anim. Sci. 11:* 599-604.

[25] Lyne, A. G.; Hollis, D. W.; Chase, H. B. 1967: Changes experimentally produced in the pigmentation of the skin and coat of sheep. *Aust. J. Sci. 30:* 30-31.

[26] Marston, H. R. 1952: Trace elements in nutrition. *Physiol. Rev. 32:* 66-121.

[27] McGavin, M. D.; Cornelius, C. E.; Gronwall, R. R. 1972: Lesions in Southdown sheep with hereditary hyperbilirubinemia. *Vet. Path. 9:* 142-151.

[28] Rae, A. L. 1956: The genetics of the sheep. *Adv. Genet. 8:* 189-265.

[29] Raper, H. S. 1928: The aerobic oxidases. *Physiol. Rev. 8:* 245-282.

[30] Ryder, M. L.; Land, R. B.; Ditchburn, R. 1974: Coat colour inheritance in Soay, Orkney and Shetland sheep. *J. Zool. Lond. 173:* 477-485.

[31] Searle, A. G. 1968: *Comparative genetics of coat colour in mammals.* Logos Press, London — New York.

[32] Shelton, M. 1966: A recurrence of the Ancon dwarf in Merino sheep. *J. Hered. 59:* 267-268.

[33] Warwick, B. L.; Jones, J. M.; Dameron, W. H.; Dunkle, P. B. 1933: Polled Rambouillet breeding *Proc. Am. Soc. Anim. Prod.:* 287-290.

[34] Wickham, G. A.; Rae, A. L. 1977: Investigations of carpet wool genes in sheep. *Proc. N.Z. Soc. Anim. Prod. 37:* 213-217.

4
Selection and its Effects

A. L. Rae

Massey University, Palmerston North
Consultants: A. H. Carter, D. C. Dalton and K. J. Dunlop

SUMMARY

This chapter aims to give a basic understanding of the basis of selection and the use of selection in the improvement of sheep flocks.

There is some discussion on the changes in gene frequency following selection for a simply inherited trait but the main emphasis is on selection where the effects of individual genes cannot be recognized and where it is difficult to separate the effects of genetic and environmental factors.

Methods of carrying out selection in sheep flocks and the relative accuracy of various selection procedures in different situations are discussed.

INTRODUCTION

For the sheep farmer selection has two types of effects. Firstly, it can result in the elimination of sheep whose presence might adversely affect the profitability of the flock because of aspects of their own performance. Secondly, selection can result in genetic changes in the level of production of sheep born in the flock. This second type of effect is relatively permanent and the changes accumulate while the first type of effect is temporary. In a typical New Zealand flock, ewe selection has both types of effects but ram selection has predominantly the second type of effect. This chapter will concentrate on the second type of effect since, in the long term, this is more important.

Selection is a process that results in one kind of individual leaving more progeny, on average, than individuals of another kind.[13] [15] The fundamental effect of this is to increase the frequency of genes that tend to result in animals with favoured phenotypes.

Thus if a farmer has some black lambs (aa, see p. 55) turn up in a white flock, where most of the sheep will be $A^{wh}A^{wh}$, he can develop a black sub-flock or even convert his whole flock to black by selection. Initially the proportion of a genes in the flock is low (gene frequency close to 0). If only black rams were used, the frequency of the a gene in the progeny would be more than 0.5. Most of the progeny of white ewes would be white but would be heterozygous for the black gene ($A^{wh}a$). In the next generation, if black rams are used again, more than half the lambs would be black and the frequency of the a gene would have risen to more than 0.75. At this stage the breeder would be able to select black ewes and cull the white ewes and this would result in the virtual elimination of the last A^{wh} genes from the flock. When all the ewes and rams in the flock were black the frequency of the a gene would have risen to 1 while the frequency of the A^{wh} gene would have fallen from just less than 1 to 0.

Several factors change the frequency of genes but selection is generally the most potent factor. The various living species have evolved largely as a result of natural selection reducing the frequency of genes that limit the ability to survive in certain environments. *Natural selection* is still an important force changing gene frequency in sheep. However, careful flock management tends to limit its impact. Thus, if there are genetic differences between sheep in susceptibility to a disease which kills or reduces the number of progeny, natural selection will lead to reduced susceptibility; effective disease treatment, if not associated with increased culling of affected animals, will prevent the genetic change. Some sheep breeders refrain from assisting ewes at lambing in the belief that the loss of ewes and lambs with lambing difficulties will result in the evolution of a flock with less lambing difficulty. When careful sheep husbandry has reduced the amount of natural selection, artificial selection becomes far more important.

Most of the economically important traits in sheep are controlled by many genes. Usually the presence of individual genes cannot be recognized and it is not possible to study changes in their frequency. However, it is clear that selection changes the frequency of genes that result in the preferred levels of traits. In this situation the effects of selection are described in terms of changes in the level of traits from one generation to the next.

THE NATURE OF VARIATION IN PRODUCTIVE TRAITS OF SHEEP

Consider first some general observations about the variation in a character such as fleece weight in a flock of sheep.

(1) There are marked differences between sheep of the same age kept within the same flock under similar conditions; e.g., a range in hogget fleece weights from about 2.0 kg to 5.0 kg would be expected in a flock where the average hogget fleece weight is 3.5 kg.

(2) Levels of feeding, management, climatic effects and other environmental factors are known to influence fleece weight. In many of the later chapters of this series much information on non-genetic effects on productive traits of the sheep will be presented.

(3) Detailed examination of fleece weight records in a flock shows

that there is some degree of likeness between related sheep in their fleece weights. For example, there is a small correlation between the fleece weights of dams and their offspring. The variation found within a group of ewes all sired by one ram is not as great as that within a group of unrelated ewes.

These facts suggest two conclusions:

(1) That environmental differences are a major source of variation in the trait.

(2) That some of the variation between animals can be ascribed to differences in their genotype. For a particular trait such as fleece weight, there is some large but unknown number of gene pairs which control its inheritance. Many of these will be in the fixed or homozygous state in the population and will contribute to the mean fleece weight of the population but not to the variation about that mean. The remaining genes will be in the unfixed state and these will contribute to the genotypic variation in the population.

Hence as a first approximation, we can say that:

Each fleece weight = Flock mean + genotypic component + environmental component.

Then, since the mean is a constant, one can state:

Phenotypic variance = Genotypic variance + Environmental variance.

This expression will be true providing that there is no correlation between genotype and environmental effects. Such a correlation could be produced in a flock if the manager sets out to give better than average environment to those sheep which are genetically better than average. An example of this would be where a breeder selects out a special group of what he thinks are his best ram hoggets and feeds them better than the rest of the ram hoggets, a not uncommon practice in stud breeding.

Subdivision of Genetic Variation

The word *genotype* has been used to refer to the whole set of genes for a given trait in the individual. This set of genes functions as a unit in the development of the individual's *phenotype*. But it does not act as a unit in transmission to the individual's progeny since it is broken up by segregation in the formation of gametes. Thus parents pass on only *half of their genes* (one of each allelic pair) not their genotypes to their progeny. Because, in selection, one wishes to choose those individuals as parents which will have the best progeny, it is necessary to have some measure which describes the actual transmission of the genes to the progeny.

As shown in Chapter 3 the way genes interact is complex. In the simplest case they combine additively; i.e., substituting one gene for another will have exactly the same effect regardless of the other genes present. Alternatively the response will depend on the other allele present *(dominance)* or on non-allelic genes *(epistasis)*. The genes controlling a trait may act in all of these ways.

Thus the effects of a gene vary in different animals. The average effect of a gene is the average change in a trait resulting from each substitution of the gene for an allele.[1] [13] [15] [17] This applies to a particular population (breed, flock, etc.). The *breeding value* of an animal is the sum of the average effect of all its genes. This is not the same as the *genotypic value;* the difference is due to dominance and epistatic effects.

Thus when many genes affect a trait the genotype can be divided into:

(1) a *breeding value* which results from simply adding up the average effects of all the genes affecting the trait.

(2) a *dominance effect* which results whenever the heterozygote is not midway between the two homozygotes. These effects are summed over all loci affecting the trait.

(3) an *epistatic effect* which results from the possibility that two or more genes may have a joint effect or may act in combination. Thus we may write:

genotype = breeding value + dominance effect + epistatic effect

or

$$h = g + d + i$$

Then $\quad \text{var } h = \text{var } g + \text{var } d + \text{var } i$

In selection, the breeding value for the trait and its associated variation is of most interest simply because it gives a prediction of the average of the individual's progeny. Dominance and epistatic deviations serve to obscure the true breeding value of the individual and make its assessment less accurate. These non-additive sources of genotypic variation are of substantially greater significance in the study of crossbreeding and inbreeding systems (see Chapter 5).

Environmental Variation

The environmental variation includes all of the variation in the trait which is non-genetic in origin. In sheep, two obvious environmental causes of variation are differences in level of nutrition and climatic effects which may affect the trait directly but, more importantly, may affect the level of nutrition in a grassland system of farming. Disease and accident are other environmental effects. Errors in the measurement of the trait also contribute. They are usually not important in traits which can be measured objectively but they can be important in traits which are assessed by eye or hand. Even after account has been taken of all known sources of variation, there is usually some remaining variation of which the cause is unknown.

Environmental variation makes the assessment of an animal's genotype less accurate but, apart from this, does not affect the progeny.

Thus feeding a ram at a high level to get large size and heavy fleece weights will have no effect on the progeny.

In traits such as fleece weight and number of lambs born (or reared) which can be measured from year to year on the same animal, it is convenient to subdivide the environmental variation into two parts:

(1) A part which is common to all of the records of the particular sheep but differs from sheep to sheep. This is referred to as the *permanent environmental variation*.

(2) A part which is peculiar to the particular record of the individual sheep and differs from record to record of that individual. This is referred to as *temporary environmental variation*.

One example of a permanent environmental effect would be liver damage resulting from intake of the facial eczema toxin, sporidesmin, which may limit the fertility of a ewe over the rest of its lifetime. Fleece weight may suffer a permanent depression as a result of a sheep being limited in size due to malnutrition early in its lifetime. In addition fleece weight may suffer a temporary effect due to level of nutrition in the year the fleece is being grown.

Heritability

The description of the measurement of a trait such as fleece weight which is repeated from year to year is:

Fleece weight = mean + breeding value + dominance effect + epistatic effect + permanent environmental effect + temporary environmental effect

or, in general for a trait x

$$x = m + g + d + i + e_p + e_r$$

It is frequently assumed that dominance and epistatic effects are small. Also, in traits that are only measured once, no distinction can be made between permanent and temporary environmental variation. Thus the model simplifies to:

$$x = m + g + e$$

In terms of the variances (measures of variation), the total observed variance can be subdivided into the variance in breeding values and the environmental variance, e.g.

$$\text{var } x = \text{var } g + \text{var } e$$

This leads to the concept of the *heritability* of a trait

$$h^2 = \frac{\text{var } g}{\text{var } x}$$

This value is extremely important in prediction of rates of response from various types of selection and these predictions are necessary when comparing alternative breeding plans.

Table 4.1 illustrates the concept that a record (x) of each sheep is the result of the overall mean plus a breeding value (g) and an environmen-

64

tal component (e). Thus for sheep 1

fleece weight = 4.0 + (−0.4) + 0.1 = 3.7

Note the way the environmental component can mask the breeding value. Thus sheep 10 is one of the top two on breeding value but its record is below average.

TABLE 4.1: Hogget Fleece Weight Records (x) Comprising a Breeding Value (g) and Environmental Component (e) in Kg (average is 4.0 kg)

Sheep	Breeding Value (g)	Environmental component (e)	Record (x)
1	− 0.4	+ 0.1	3.7
2	+ 0.2	− 0.1	4.1
3	+ 0.2	− 0.2	4.0
4	0	+ 0.3	4.3
5	− 0.2	+ 0.1	3.9
6	− 0.3	+ 0.2	3.9
7	− 0.2	− 0.7	3.1
8	+ 0.3	+ 0.5	4.8
9	+ 0.1	+ 0.3	4.4
10	+ 0.3	− 0.5	3.8
$\Sigma(x_i - \bar{x})^2$	0.60	1.28	1.86
Variance	0.066	0.140	0.206

The variances of the trait, its breeding value and the environmental component are also presented in Table 4.1. Because the breeding values and environmental components are expressed as deviations from the mean, they need only be squared, summed and divided by 9 to obtain the variance. The relationship (where there is no correlation between g and e) is then

$$\text{var } x = \text{var } g + \text{var } e$$
$$\text{or } 0.206 = 0.066 + 0.140$$

$$\text{heritability } (h^2) = \frac{\text{var } g}{\text{var } x} = \frac{.066}{.206} = 0.32$$

The heritability can also be shown to equal the regression of the breeding value on record:

$$h^2 = b_{gx} = \frac{\Sigma(g-\bar{g})\,(x-\bar{x})}{\Sigma(x-\bar{x})^2} = \frac{0.59}{1.86} = 0.32$$

and also the square of the correlation between the breeding value and the record:

$$h^2 = r_{gx}^2 = \left(\frac{\Sigma(g-\bar{g})\,(x-\bar{x})}{\sqrt{\Sigma(g-\bar{g})^2\,(x-\bar{x})^2}} \right)^2$$

$$= \left(\frac{0.59}{\sqrt{0.60 \times 1.86}} \right)^2 = 0.56^2 = 0.32$$

Unfortunately the heritability cannot be calculated as above because the true breeding values are not known. It can be estimated by studying the degree of resemblance between related animals.[13][15]

There is a general tendency for parents with above average fleece weights to have offspring that are also above average. Each daughter whose fleece weight record is plotted in Fig. 4.1 will have received half the genes carried by her dam. The correlation between the dams' and the daughters' records is a measure of the relationship between them. On average, half each dam's breeding value for fleece weight will have been transmitted to its daughter. Therefore doubling the correlation between fleece weights of the dams and daughters will give an estimate of the heritability of fleece weight. This estimate will contain a little of the epistatic variance if any is present, but none of the dominance variance. It may include some extra likeness between the dam and her offspring because the dam's milking and mothering ability can influence the offspring (maternal effects). This is especially likely to occur in measures of early growth rate, e.g., weaning weight.

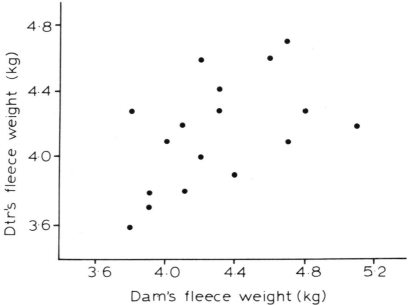

Fig. 4.1: Relationship between fleece weight of dam and daughter.

The other relationship which is commonly used for estimating heritability is that between paternal half-sibs (i.e. between individuals which are by the same sire out of different dams). Paternal half-sibs have one-quarter of their genes in common; hence multiplying the correlation between paternal half-sibs by four gives an estimate of heritability. It may contain a little of the epistatic variance but no dominance variance and no maternal effects. Other relationships are not often used in estimating heritabilities in sheep.

Some estimates of heritabiity of traits of economic importance in New Zealand breeds of sheep are listed in Table 4.2. It is important to note that heritability is a property of a particular trait in a specified population. The variation both in breeding values and in environmental conditions may differ from one population to another. More variable environmental conditions will reduce heritability; more uniform conditions increase it.

For some purposes (mostly concerned with the rate of response of the trait to selection), it is convenient to classify heritabilities of 0.3 and above as high; 0.1 to 0.3 as intermediate and those below 0.1 as low.

TABLE 4.2: The Most Common Range of Reliable Estimates of Heritability and Repeatability for New Zealand Romneys and Perendales plus Australian Corriedales and Merinos

Trait	Romney	Perendale	Corriedale	Merino
Heritabilities				
Greasy fleece weight	0.3-0.4	0.3-0.4	0.2-0.4	0.3-0.4
Clean fleece weight			0.3-0.4	0.3-0.5
Clean scoured yield	0.2-0.5		0.4-0.5	0.4-0.5
Staple length	0.3-0.6	0.3-0.6	0.4-0.6	0.3-0.6
Mean fibre diameter	0.4-0.7	0.4-0.6	0.4-0.6	0.4-0.6
Fibre diameter variation	0.3-0.6			
Quality number	0.3-0.5	0.2-0.3	0.3-0.5	0.3-0.4
Crimps per unit length	0.5-0.7		0.2-0.3	0.4-0.6
Total crimp number	0.7-0.8			
Fleece character	0.1-0.3	0.1-0.4	0.3-0.5	0.3-0.5
Tippiness	0.1-0.3			
Medullameter index	0.4-0.8			
Handle	0.2-0.5		0.1-0.3	0.1-0.3
Lustre	0.3-0.4			
Greasy colour (midside)	0.2-0.4		0.3-0.4	0.3-0.4
Scoured colour	0.1-0.2			0.2
Discoloured area	0.1-0.2			
Cotting (midside)	0.1-0.4			
Cotting (fleece)	0.1-0.4			
Soundness	0.1-0.2			
Weaning weight	0.1-0.4	0.2		0.2-0.3
Hogget liveweight	0.2-0.5	0.3-0.4		0.4-0.6
Lambs born (2 yr)	0.0-0.1			0.0-0.2
(2-4 yr)	0.0-0.3			
Lambs weaned (2 yr)	0.0-0.1	0.0-0.2		0.0-0.1
(older)		0.0-0.2		0.0-0.2
Hogget oestrus	0.3-0.5			
Repeatabilities				
Greasy fleece weight	0.4-0.7	0.5-0.7	0.4-0.8	0.4-0.8
Staple length	0.4-0.7	0.4-0.6	0.4-0.8	0.4-0.8
Mean fibre diameter		0.5-0.9		0.4-0.7
Lambs born	0.1-0.25			0.01-0.1
Lambs weaned	0.1-0.25	0.05-0.2		0.05-0.1

References: 2, 3, 7, 9, 10, 12, 18, 23, 26.

One generalisation which seems to hold with few exceptions is that traits of the animal which are closely connected with natural or reproductive fitness (often measured by the number of offspring which the individual contributes to the next generation) have low heritabilities.[13] Thus one would expect heritability of number of lambs born or reared (which is likely to be closely related to fitness) to be lowly heritable while fleece traits which generally would have a lower correlation with fitness to have intermediate to high heritabilities.

Repeatability

In traits such as fleece weight and number of lambs born, which are repeated from year-to-year, repeatability is defined as the correlation between the records of the same animal. It is thus a measure of the similarity between successive records of the same animal and equals

$$t = \frac{var\ g + var\ e_p}{var\ x}$$

As with heritability, it may vary from one trait to another and from population to population. It is used mainly in deciding to what extent averaging several measurements of the same animal may increase the effectiveness of selection. It is also a measure of how much gain can be made during the lifetime of the selected animals in the flock.

An important assumption involved in repeatability is that the measurements taken are in fact controlled by the same breeding value, i.e., that the genes which control say hogget fleece weight are the same as those controlling two-tooth fleece weight. Estimates of repeatability for some traits in New Zealand sheep are listed in Table 4.2.

Genotype x Environment Interactions

To simplify the explanation so far it has been assumed that the genotype (breeding value) and environmental effects add together to give the phenotype. However, genotype x environment interactions (GEI) frequently complicate matters. If the relative performance of one genotype compared with another changes as the environment changes then GEI is said to exist. For example a group of Romney sheep will probably have heavier fleece weights than Merinos if both were grazing lush flat pastures in a wet region. However, on dry mountainous tussock pastures the Merinos would probably have higher fleece weights than the Romneys.

A recent trial comparing hoggets, sired by the same rams, reared at different stocking rates indicated that GEI were not major causes of variation for wool traits but they had a profound influence on spring live weight of hoggets.[3] Table 4.3 shows the average live weights of daughters of seven rams. These were reared at 16 and 26 ewe equivalents per hectare and are part of one year's results from this trial. It can be seen that the ranking of the sire groups changed dramatically with stocking rate. For example, the daughters of sire 6 ranked best when reared at a high stocking rate but were poorest at the lower stocking rate.

TABLE 4.3: Mean Hogget Live weight (Kg) and Ranking within each stocking rate of daughters of different sires

Sire	16 ewes/ha		26 ewes/ha	
	Weight	Rank	Weight	Rank
1	37.1	6	37.3	7
2	39.2	2	37.7	5
3	38.7	3	37.8	4
4	37.5	5	38.1	3
5	37.8	4	38.3	2
6	36.3	7	39.3	1
7	39.3	1	37.5	6

Heritabilities are normally calculated within specific environments. If genotype x environment interactions exist and sheep (rams) are selected in one environment and then their progeny have to perform in another, the selection becomes less accurate than the heritability would suggest. Thus if a farmer operating at a high stocking rate wishes to buy rams that will improve his hogget live weight, he is likely to make more progress when selecting rams reared at a high stocking rate than if he selects rams reared at a low stocking rate (even if the rams reared at the high stocking rate are smaller than those from the low stocking rate).

AN EXAMPLE OF CULLING AND SELECTION

Systems of artificial selection in sheep vary tremendously. Frequently the criteria are subjective with the sheep being selected or culled solely on the basis of a combination of visual characteristics. These subjective systems are not usually efficient methods of selecting more productive animals and also the effects are very difficult to predict. They tend to be greatly influenced by aesthetic values which are often not stable and are difficult to quantify. In comparison most modern systems of sheep selection are largely based on objectively-assessed criteria where concise written or electronically-stored records can be kept and where the response to selection in a flock can usually be assessed fairly accurately.

For these reasons the example of selection chosen is a study of the predicted response to selection for higher fleece weight in a flock where 20 ewe and 20 ram hoggets were available as replacements and 14 ewes and 2 rams are needed. Column 4 of Table 4.4 gives the fleece weights (x) of 20 sheep. The average is 4.0 kg. The fleece weights of 14 sheep left after 30% had been culled are shown in column 5. The difference between the original average (4.0 kg) and the average of the selected hoggets (4.19 kg) is (4.19 − 4.0) = 0.19 kg. This value, the *selection differential,* is a measure of the amount of selection. A selection differential of about 0.19 kg is fairly typical of what can be achieved if a flock of ewe hoggets is culled solely on fleece weight.

TABLE 4.4: Data Illustrating Selection for Hogget Fleece Weight (Kg) (Average is 4.0 Kg, Heritability 0.34 and Standard Deviation 0.45 Kg)

(1) Tag No.	(2) g_d	(3) e	(4) x	(5)	(6) g_s	(7) $\dfrac{g_d + g_s}{2}$	(8) Segrn Effect	(9) g_o	(10) e_o	(11) x_o
1	0.1	0.2	4.3	4.3	0.3	0.20	0	0.2	0.3	4.5
2	−0.4	0.3	3.9	3.9	0.4	0	−0.20	−0.2	0	3.8
3	0.2	−0.7	3.5	—						
4	0	0.1	4.1	4.1	0.3	0.15	−0.05	0.1	0.1	4.2
5	−0.1	0.3	4.2	4.2	0.4	0.15	0.21	0.4	−0.4	4.0
6	0.4	0.2	4.6	4.6	0.3	0.35	0.20	0.6	0	4.6
7	0.1	−1.1	3.0	—						
8	−0.3	0.3	4.0	4.0	0.4	0.05	0.14	0.2	−0.1	4.1
9	−0.1	−0.2	3.7	—						
10	`0.1	−0.4	3.7	—						
11	0.3	0	4.3	4.3	0.3	0.30	−0.20	0.1	0.3	4.4
12	0	−0.1	3.9	3.9	0.4	0.20	−0.11	0.1	−0.4	3.7
13	0.1	1.0	5.1	5.1	0.3	0.20	−0.08	0.1	0.1	4.2
14	−0.3	0.5	4.2	4.2	0.4	0.05	0.27	0.3	0.3	4.6
15	0.1	0.3	4.4	4.4	0.3	0.20	−0.09	0.1	−0.2	3.9
16	0	−0.2	3.8	3.8	0.4	0.20	0.14	0.3	0	4.3
17	−0.2	0	3.8	3.8	0.3	0.05	−0.20	−0.1	−0.3	3.6
18	0.1	0	4.1	4.1	0.4	0.25	−0.11	0.1	0.2	4.3
19	0	−0.3	3.7	—						
20	−0.1	−0.2	3.7	—						
Sum			80.0	58.7						58.2
Average			4.0	4.19						4.16

Selection differential (ewes) 4.19 − 4.0 = 0.19 kg
(rams) 4.75 − 4.0 = 0.75 kg
Average selection differential $\dfrac{0.19 + 0.75}{2}$ = 0.47 kg

Genetic gain = 4.16 − 4.00 = 0.16
Realised heritability = 0.16/0.47 = 0.34

Relatively few rams are required and two might be sufficient to retain for breeding. If the rams had the same distribution of fleece weights shown in column 4 the two selected would have an average fleece weight of

$$\frac{4.8 + 4.7}{2} = 4.75 \text{ kg}$$

In this case the selection differential would be (4.75 − 4.00) = 0.75 kg. Since only 10% of the rams were needed compared with 70% of the ewes a considerably higher selection differential has been achieved.

One of the rams has a breeding value of + 0.3, the other + 0.4. Column 6 (g_s) shows the breeding value of the ram used with each ewe

at 2 year-old mating. The hogget fleece weights of the offspring are listed in column 11. The average fleece weight of the offspring is
$$58.2 \div 14 = 4.16 \text{ kg.}$$
The gain achieved or the response to selection is then measured as the difference between the mean of the unculled offspring and the mean of the unculled parental generation, i.e.,
$$(4.16 - 4.0) = 0.16 \text{ kg.}$$
Thus, although the parents themselves averaged
$$\frac{0.19 + 0.75}{2} = 0.47 \text{ kg}$$
above the original mean of 4.0 kg, their offspring only average 0.16 kg above the mean. The ratio of gain/selection differential can be thought of as a measure of the efficiency of the selection process since it represents what was achieved as a proportion of what was attempted. In fact, this ratio is equal to the heritability of fleece weight,[13] i.e.

$$\frac{\text{Genetic gain}}{\text{Selection differential}} = \text{Heritability}$$

and, by rearrangement

genetic gain = heritability × selection differential

Table 4.4 contains information which helps in understanding the inheritance of a quantitative trait such as fleece weight. In columns 2 and 3 are respectively the breeding value and environmental component for each animal, which when added to the general mean (4.0 kg) gives the fleece weight in column 4 (see also Table 4.1). In column 6 as indicated before, the breeding value of the two rams used is listed in relation to the ewes to which each was mated. Because the breeding values are the sum of the average effects of genes affecting the trait, the expected breeding value of the offspring is simply midway between the breeding value of its two parents;

$$\text{i.e., } g_o = \frac{g_s + g_d}{2}$$

or in the case of sheep No. 1,

$$\frac{0.3 + 0.1}{2} = 0.20 \text{ kg.}$$

This expected value controls one-half of the variance in breeding value of the offspring, the remaining half being controlled by the random assortment of the genes in Mendelian segregation. This latter part is shown in column 8. The sum of column 7 and column 8 then specify the breeding value of the offspring (g_o) in column 9. A random environmental component (e_o) in column 10 is added to give the fleece weight of the offspring (x_o) in column 11.

From this table, it is noted that the breeding values of the parents are important elements in controlling the record of the progeny. This is because the breeding value of the progeny is expected to lie mid-way between the breeding values of the parents. The effect resulting from the random or chance segregation of the parents' genes is not under control

71

of the breeder. However, it tends to average zero over a number of offspring of the same mating as indicated in Table 4.5. Thus the average breeding value of a large number of progeny would be exactly mid-way between those of the parents. Further, the environmental component will average zero over a large number of progeny. Hence:

$$\text{Average of progeny} = \text{mean} + \frac{g_s + g_d}{2} + \text{average segregation effect}$$
$$+ \text{ average environmental effect}$$

Since the last two will approach zero, the average of the progeny of a mating expressed as a deviation from the flock average will equal the average breeding value of the parents.

TABLE 4.5: The Relationship between Breeding Values of Offspring and the Breeding Values of Parents

Offspring	Breeding value sire (g_s)	Breeding value dam (g_d)	$\frac{g_s + g_d}{2}$	Segrega-tion effect	Breeding value (g_o)
1	0.3	0.1	0.2	+0.2	0.4
2	0.3	0.1	0.2	−0.1	0.1
3	0.3	0.1	0.2	0	0.2
4	0.3	0.1	0.2	−0.2	0
5	0.3	0.1	0.2	+0.1	0.3
Sum				0	1.0
Average					0.2

When referring to the breeding value of an individual, it is assumed that it is mated to a large number of the opposite sex in the flock and consequently the breeding value of these mates will average zero. Then, for example, the average breeding value of the progeny of a sire will be:

$$\frac{g_s + 0}{2} = g_o$$

Thus if the breeding value of a ram for fleece weight is stated to be 0.2 kg, this implies that, if the ram was mated to a large number of ewes in the flock, then the offspring would average ($\frac{1}{2} \times 0.2$) kg = 0.1 kg above the average fleece weight of the flock.

In selection for genetic improvement, the breeder is concerned with the question as to how well the offspring of each individual are likely to perform on average in the flock. This is exactly the information supplied by each individual's breeding value.

Breeding values cannot be measured directly or be known exactly as has been assumed for illustrative purposes in the present discussion. They can only be estimated from records on the individual animal or its relatives. It is thus necessary to distinguish between the true breeding value (g) and estimates of this derived from records ($\hat{g}$).

Successful selection depends on doing two steps efficiently:

(1) Predicting the breeding value of the individual animal for the trait or traits to be selected for.

(2) Deciding on the basis of the predicted breeding value whether the animal should be kept or culled and then taking the appropriate action.

SYSTEMS OF SELECTION

Selection Based on a Single Record of each Individual

The problem is to predict the breeding value of an individual from a single record. This is accomplished by the following computation:

Breeding value (estimated as a deviation from the mean of the flock)
= heritability × deviation of the record from its mean.

Thus using hogget fleece weight with a heritability of 0.30 as an example, then for sheep No. 1 in Table 4.4

$$\text{Estimated breeding value} = 0.30 \ (4.3 - 4.0)$$
$$= 0.30 \times 0.30$$
$$= +.09 \text{ kg}$$

When only a single record is used, it can be seen that the predicted breeding values will rank the sheep within the flock in exactly the same order as their actual fleece weights. Thus, in these circumstances, the decisions as to which sheep are to be selected can be made on the actual fleece weights as was done in Table 4.4. For selection based on a single record:

genetic gain per generation = heritability × selection differential

The relationships above show the importance of the heritability of a trait:

(1) It allows the prediction of the breeding value of the individual. As noted on p. 65, it is the regression coefficient of breeding value on the phenotypic record (b_{gx}).

(2) It allows the prediction of the gain which can be achieved by selection. If the heritability of the trait is high, gain per generation will be rapid. Where it is low, gain through individual selection will be slow.

(3) When heritability is high, the phenotypic record is an accurate predictor of the breeding value of the individual. This can be seen by noting, as on p. 65, that the correlation between breeding value and phenotypic record is $\sqrt{\text{heritability}}$. The higher the correlation between the two variates, the more accurate is the prediction of breeding value.

The Selection Differential

The genetic gain from selection depends on the size of the selection differential. This, in turn depends very much on the proportion of the sheep available that are required for breeding. As seen previously fewer rams are required for breeding than ewes and therefore a higher selection differential can be achieved with rams. The selection differential for rising two-tooth ewes can be increased either by increasing the number available or reducing the number required for breeding.

Factors which increase the number of two-tooth ewes available for selection are:
(1) High lambing percentages. These result in a higher number of offspring available at birth or docking.
(2) Low mortality rates in lambs after birth or docking through to the time selection is made.

Factors which reduce the number of replacements required to maintain flock numbers are:
(1) Low mortality and culling rates in the ewe flock.
(2) The older the ewes are at casting-for-age, the smaller the number of replacements needed.
(3) Reduction in flock size.

If the trait which is under selection follows a normal distribution, it is possible to determine the selection differential from knowledge of the proportion of the flock selected and the standard deviation of the trait. Table 4.6 shows the selection differential for a normally-distributed trait with a standard deviation of one unit. The use of the table can be shown by the following example. Given that the standard deviation of hogget fleece weight is 0.45 kg and that the best 60% of the ewe hoggets and 3% of the ram hoggets are selected on fleece weight, then:

Selection differential (rams) 2.27 × 0.45 kg = 1.02 kg.
Selection differential (ewes) 0.64 × 0.45 kg = 0.29 kg.

The selection differential as calculated above can be used especially to give a general indication of the effects of different selection plans.

Generation Interval

In sheep a high proportion of the total selection operates only once in the lifetime of each animal. That is, once an animal has been selected to enter the flock it will tend to stay in until it dies or is cast-for-age. Consequently the genetic gain discussed earlier is the amount of gain achieved over a generation. It is often required to know the amount of genetic gain per year. This is given by:

$$\text{Genetic gain/year} = \frac{\text{Genetic gain/generation}}{\text{Generation interval}}$$

The generation interval is defined as the average age of the parents when their offspring are born. An example of the calculation for a flock of 100 sheep is shown in Table 4.7

The generation interval can be changed by the breeding and management systems used in the flock. For example, if the breeder uses

TABLE 4.6: Selection Differentials for a Trait with Standard Deviation of One Unit

Proportion selected	Selection differential
0.01	2.66
0.02	2.42
0.03	2.27
0.04	2.15
0.05	2.06
0.10	1.76
0.15	1.55
0.20	1.40
0.25	1.27
0.30	1.16
0.35	1.06
0.40	0.97
0.45	0.88
0.50	0.80
0.55	0.72
0.60	0.64
0.65	0.57
0.70	0.50
0.75	0.42
0.80	0.35
0.85	0.27
0.90	0.20
0.95	0.11

TABLE 4.7: Calculation of the Generation Interval for Ewes in a 100 Ewe Flock

Age of ewe at birth of lamb (1)	No. of ewes (2)	No. of lambs/ewe (3)	No. of lambs (2) x (3)	Product lambs x years (1) x (2) x (3)
2 years	29	0.9	21	42
3 years	27	1.1	30	90
4 years	23	1.3	30	120
5 years	21	1.4	29	145
All ages	100		110	397

Generation Interval $= \dfrac{397}{110} = 3.61$ years

two-tooth rams only once in the flock and replaces them each year with a new batch of two-tooth rams, then the generation interval for rams is two years. Generally in sheep the generation interval on the ewe side will vary between 3½-4½ years and can only be changed a little because a comparatively high number of ewe replacements are required. The generation interval on the sire side may vary more widely from about 1 year to 3½ years.

The generation interval is important in comparing the results of different kinds of selection. For example, the use of progeny testing (because it is necessary to wait to obtain the records of the progeny of a sire) will often increase the generation interval. But although progeny testing will usually give a better genetic gain per generation than individual selection, the increase in the generation interval may offset this to result in a lower genetic gain per year.

Adjusting Records for Non-Genetic Effects

The effect of the environment in masking the breeding value of the individual sheep has been stressed already. The environmental component is usually very complex but in some cases it is possible to identify some specific effects. For example, it is known that twins have a lighter weaning weight than single-born lambs because they have shared their pre- and post-natal maternal environment. But, if it is known that twins are on average 4.0 kg lighter than singles and if this difference is added to the weaning weight of the twin lambs, then the effect of the difference between twins and singles is eliminated from the comparison between them. Thus the variation caused by the twin-single difference is taken out of the environmental variance of weaning weight. This results in the heritability of the *adjusted weaning weight* being higher than the heritability of the unadjusted weight and a greater genetic gain from selection can be expected.

TABLE 4.8: Correction Factors for Weaning Weight (Dual Purpose Breeds)

Factor		Correction (kg)
Birth and Rearing Bank		
Born as	**Reared as**	
Single	Single	0
Twin	Single	+2.0
Twin	Twin	+4.2
Age of Dam		
2 year		+1.3
3 year		+0.2
4 year +		0
Age of lamb at weaning		
For each day **younger** than average		+0.17
For each day **older** than average		−0.17

Correction factors for non-genetic effects on weaning weight for dual-purpose breeds (as used in Sheeplan) are presented in Table 4.8 along with an assessment of the percentage of the total variation in weaning weight controlled by each source of variation.[7] The use of these correction factors is illustrated in Table 4.9. For example, lamb 3/76 is a twin reared as a twin. Hence 4.2 kg is added to adjust it to a single. Its dam was a 3-year-old ewe, hence 0.2 kg is added. It is 2 days older than the average age at weaning; so 2 × 0.17 = 0.34 kg is subtracted from actual weight. These adjustments total (4.2 + 0.2 − 0.34) kg or + 4.06 kg. This is added to the actual weaning weight of 22 kg to give 26.1 kg as the adjusted weight.

TABLE 4.9: Example of Adjusting Weaning Weight for Environmental Effects

Tag No.	Birth + Rearing Rank	Age of Dam (year)	Age of Lamb (days)	Actual wt (kg)	Adjusted wt (kg)
3/76	2/2 (+4.2)	3 (+0.2)	+2 (−0.34)	22 (+4.1)	26.1
4/76	2/2 (+4.2)	3 (+0.2)	+2 (−0.34)	18 (+4.1)	22.1
6/76	2/1 (+2.0)	2 (+1.3)	−3 (+0.51)	24 (+3.8)	27.8

Another technique used to eliminate the influence of a non-genetic factor is to express each record as a *deviation* from the average of the group to which it belongs. For example, it is possible to remove the effect of sex on weaning weight by calculating the average for the ram lambs and expressing each ram lamb's weaning weight as a deviation from the average. Similarly, the weaning weight of each ewe lamb is expressed as a deviation from the average of all ewe lambs. This method is used when the effects of the environmental factor are variable from flock to flock and from year to year. An example of the use of this method is in adjusting for age-of-ewe and year effects on number of lambs born as used in Sheeplan (Table 4.10).[7]

TABLE 4.10: Expressing Number of Lambs Born (NLB) as a Deviation from Year-Age Average

Ewe No.	Year of Record	NLB	Year-age Average	Deviation
5/72	1974	1	0.80[a]	+0.20[b]
	1975	1	1.00	0
	1976	2	1.10	+0.90

[a] 0.80 is the average NLB of all two-tooth ewes in the flock in 1974. 1.00 is the average of all four-tooth ewes in 1975, etc.
[b] The average deviation for NLB is 1.1/3 = 0.37 lamb

Sometimes, environmental variation can be controlled by physical standardisation so that all animals in the flock have the same treatment. For example, with cattle growth performance tests under intensive conditions, a standard ration is commonly used. However, with sheep under pasture conditions, physical standardisation of the environment can seldom be attained.

Selection Based on the Average of Several Records

With traits such as fleece weight and number of lambs born or reared where the individual sheep can have several records, the average of the records can be used as the basis for predicting the breeding value.

Using the average of several records makes it possible for the temporary environmental effects peculiar to each record to cancel each other out, at least in part. In an average based on k records, only 1/k of the temporary environmental variance will remain. Thus the phenotypic variation in the trait is reduced but the variation in breeding values remains unchanged. Hence the heritability of the average is higher than that of a single record; the accuracy of predicting the breeding value and the gain per generation are increased.[15]

It is possible to see how the increase in heritability of an average occurs by using the description of fleece weight given on p. 64. There it was assumed that:

Fleece weight = mean + breeding value + permanent environmental effect + temporary environmental effect

$$\text{or} \quad x = m + g + e_p + e_r$$

TABLE 4.11: The Structure of Repeated Fleece Weight Records Including a Permanent Environmental Effect (e_p) (Average is 4.0 Kg)

Tag No.	Age	Breeding value (g)	Perm. Env. (e_p)	Temp. Env. (e_r)	Record (x)
1	Hogget	−0.4	0.1	0.1	3.8
	2th	−0.4	0.1	0	3.7
	4th	−0.4	0.1	−0.2	3.5
	Sum	−1.2	0.3	−0.1	11.0
	Mean	−0.4	0.1	−0.03	3.67
4	Hogget	+0.2	−0.1	−0.1	4.0
	2th	+0.2	−0.1	+0.2	4.3
	4th	+0.2	−0.1	+0.1	4.2
	6th	+0.2	−0.1	−0.1	4.0
	Sum	+0.8	−0.4	10.1	16.5
	Mean	+0.2	−0.1	+0.03	4.11

This model is illustrated for fleece weight in Table 4.11. From this it can be seen that the averaging process does not affect the breeding value and permanent environmental effect since by definition they are the same for each record of the animal. But the temporary environmental effect is divided by the number of records averaged. Likewise, the

78

variance of the average of k records is:

$$var\ (\bar{x}) = var\ g + var\ e_p + \frac{var\ e_r}{k}$$

and thus the heritability of the average is:

$$h^2_{\bar{x}} = \frac{var\ g}{var\ g + var\ e_p + \dfrac{var\ e_r}{k}}$$

which can, by using the definitions of heritability and repeatability, be shown to equal

$$h^2_{\bar{x}} = \frac{kh^2}{1 + (k-1)\,t}$$

where t is repeatability of the trait.

Simply because the variance of the average is less than that of a single record, the heritability of the average will be higher. The extent of the increase depends on the size of the temporary environmental variance which is the only component to be reduced. This is shown in Table 4.5 (Section 2) where the advantages in using averages are greatest when repeatability is low because temporary environmental variation is large. Low repeatability implies that the correlation between successive records is small and thus the first record is not a very good indicator of the subsequent records of the sheep. On the other hand high repeatability means a close correlation between the records of the individual and the first record is a good predictor of the remainder. Hence in the case of low repeatability taking an extra record may help increase accuracy of selection whereas with high repeatability it is unlikely to do so.

A further consideration in the use of averages of several records on the same animal is that the delay involved in obtaining the extra records will usually increase the generation interval. Thus if we wish to use two records on fleece weight, it will usually mean waiting for a further year for the extra record to be obtained. Hence, while genetic gain per generation may be increased by the use of the average, the gain per year may be decreased.

The prediction of the breeding value of an individual from an average of several records is:

$$\hat{g} = \frac{kh^2}{1 + (k-1)\,t} \quad \text{(Deviation of the average from flock mean)}$$

For example in Table 4.11 Ewe 1 with 3 fleece weight records averaging 3.67 kg has

$$\hat{g} = \frac{3 \times 0.3}{1 + (2)\,0.6}\ (3.67 - 4.0)$$

$$= 0.41\,(-0.33) = -0.14\ kg$$

Ewe 4 with 4 records averaging 4.11 kg has

$$\hat{g} = \frac{4 \times 0.3}{1 + (3)\,0.6}\,(4.11 - 4.0)$$
$$= 0.43\,(0.11) = +\,0.05 \text{ kg}$$

The accuracy of prediction as measured by the correlation between breeding value and the average of the records is the square root of the heritability of the average ($\sqrt{0.41} = 0.64$ for Ewe 1 and 0.66 for Ewe 4).

Selection for Several Characteristics at the Same Time

The sheep breeder must consider at least several different characteristics in making his selection decisions. Thus, in a dual-purpose breed, he is concerned to improve not only number of lambs produced and their growth rate but also the weight and merit of the fleece. The relative economic values to be attached to each trait are discussed in Chapter 6. In this section, the genetic aspects of selection for more than one trait at a time are considered.

Correlations between Traits

In selecting for two traits, it is necessary to take into account that they may be independent of each other, or be either positively or negatively correlated.[13][17] The association which can be measured by the correlation between, say, weaning weight and hogget fleece weight, each being measured on the same sheep, is called the *phenotypic correlation*. This correlation may be the result of two causes:
(1) Some of the genes affecting weaning weight may also affect fleece weight. This gives rise to a *genetic correlation*.
(2) Weaning weight may be correlated with fleece weight because some of the environmental factors influencing one may also affect the other. For example, better feeding is likely to increase both weaning weight and hogget fleece weight. This gives rise to an *environmental correlation*.

Hence, the phenotypic correlation between two traits is partitioned into a genetic and environmental component. This partitioning is important since it is the genetic correlation between the two traits which decides what change will take place in one trait as a result of selection for the other. Hence, unless the phenotypic correlation is divided into its two parts, it is impossible to forecast the genetic consequences which selection will have in the flock.[8] Generally, a positive genetic correlation between two traits means that selecting for one will bring about an increase in the other. On the other hand, a negative genetic correlation implies that selection for one trait will cause some deterioration in the other.

The implications of a genetic correlation between two traits can be seen from the following example. Fleece weight and fineness (as indicated by quality number) are genetically correlated with finer fleeces being associated with lighter fleece weight. If one divided the ewes in the flock into two groups, those above average in fleece weight and those below average, and then examined the fineness of the fleeces of the daughters, the tendency would be for dams with above-average fleece weight to have daughters with coarser fleeces and vice-versa. If one wishes to select for heavier and finer fleeces, then the genetic correlation is negative and progress for this combination would tend to be slow.

In general, genetic correlations are produced if some genes affect more than one trait. Since hereditary differences between individuals are produced by the effects on developmental and metabolic processes, a gene which alters a particular process will have an effect on any trait which is influenced by this process. Even though the gene may have only a single primary action the subsequent effects of this action may influence many traits. One example, already noted is the multiple effects of the N^d gene of the Drysdale which produces a speciality carpet type fleece but also leads to growth of horns.

Estimating Genetic Correlations

The genetic correlation between two traits can be estimated by calculating the correlation between one trait in the parent and the other trait in the offspring. As with estimating heritability, the paternal half-sib relationship can also be used.[13] [15] [17]

Some estimates of phenotypic and genetic correlations for traits in some New Zealand breeds are presented in Table 4.12.

Methods of Selecting for Several Traits

Selection for several traits can be done in at least three general ways:[1] [13] [15]

(1) Select for one character until it has been improved to the required level. Then select for the second character, then for the third, until all have been improved. This is called *tandem selection*.

(2) Establish a level for each character and cull all individuals which fall below the level in any character. This is called the *method of independent culling levels*.

(3) Establish a total score or *selection index* which adds up the merit of the animal for each of the traits being selected.

Tandem selection is the simplest of the three methods but is rarely used in practice. In part this is because it does not normally lead to simultaneous improvement in the traits and thus a long period of time will elapse before the final combination of traits is achieved. It is also generally the least efficient of the three methods.

The *method of independent culling levels* for two traits, hogget fleece weight and body weight, is described in Fig. 4.2. Each dot represents an

TABLE 4.12: Estimates of Phenotypic (P) and Genetic Correlations (G) between some Production Traits (at 12-18 months of age)

Traits		Romney	Perendale	Merino
Greasy fleece weight	P	0.53	0.50	0.13
x Fibre diameter	G	0.58 to 0.81	0.43 to 0.44	0.13 to 0.47
x Quality number	P	−0.03 to −0.33	−0.16	
	G	−0.49 to 0.02	−0.48 to 0.09	
x Staple length	P	0.22 to 0.51	0.44	0.25 to 0.30
	G	0.21 to 0.60	0.44 to 0.76	0.13 to 0.89
x Hogget liveweight	P	0.45 to 0.61	0.39	0.24 to 0.36
	G	0.11 to 0.54	−0.07 to 0.18	−0.11 to 0.26
x Lambs born/ewe joined	P			−0.09
2 yrs	G			−0.52
First three lambings	P			0.09
	G			0.34
x Lambs weaned/ewe joined	P		−0.11	−0.12
2 yrs	G		−0.47	−0.85
First three lambings	P			0.03
	G			0.34
Fibre diameter	P	−0.63	−0.26	
x Quality number	G	−0.09	−0.27 to 0.46	
x Staple length	P	0.41 to 0.48	0.34	0.11
	G	0.41	0.31 to 0.53	−0.42 to 0.44
x Hogget liveweight	P	0.29	0.15	0.13
	G	0.02 to 0.24	−0.06 to 0	−0.21 to 0.12
Quality number	P	−0.46 to −0.69	−0.45	
x Staple length	G	−0.53 to −0.76	−0.41 to −0.63	
x Hogget liveweight	P	0.08 to 0.22	−0.03	
	G	0.37	−0.10 to 0.37	
Staple length	P	0.01 to 0.24	0.13	0.06 to 0.10
x Hogget liveweight	G	0.21 to 0.50	−0.06 to 0.22	−0.60 to 0.47
Hogget liveweight	P	0.49 to 0.62	0.44	0.65
x Weaning weight	G	0.62 to 0.90	0.13 to 0.66	0.50 to 0.93
x Lambs born/ewe joined	P	0.23		0.07
2 yrs	G	0.56 to 0.81		0.20
First three lambings	P	0.15 to 0.23		0.12
	G	−0.24 to 0.65		0.23
x Lambs weaned/ewe joined	P		0.08	0.06
2 yrs	G			0.06
First three lambings	P	0.12		0.15
	G	0.11		0.47

References: 2, 3, 7, 9, 12, 18, 24, 26

individual hogget, the height of the dot on the vertical axis being its fleece weight and its distance along the horizontal axis being its body weight. The line AA′ at fleece weight (3.45 kg) is the culling level for fleece weight and all animals falling below it are culled. Similarly BB′ is the culling level for body weight and all animals below it are culled. The method is widely used in practice, particularly in culling for faults where breeders tend to set standards for under- and over-shot jaws, breed points, fleece character, etc., and cull all animals which do not meet them. It is simple to operate, and does not require computational facilities. Also culling can be carried out at different stages of the animal's life. For example, if selection is for high weaning weight and high hogget fleece weight, then all lambs which failed to meet the weaning weight standard would be culled at weaning and only the remainder carried on to be assessed for fleece weight.

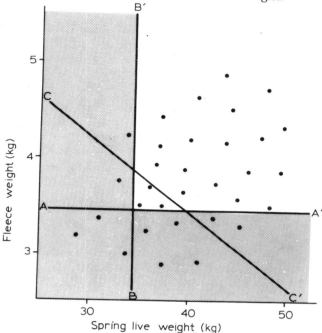

Fig. 4.2: Comparison of animals culled by independent culling levels (in shaded area) and index selection.

Fig. 4.2 also shows the important difference between the *selection index* and independent culling levels. If fleece weight and body weight are given equal emphasis then CC' represents the culling level in the selection index, all animals to the left of the line being culled. It is seen that the selection index allows high values of fleece weight to compensate to some extent for low body weight and *vice versa*.

The relative efficiency of the three methods is:[28]

(1) The index-method is never less efficient than independent culling levels. Independent culling is never less, but in some cases no more efficient than tandem selection.

(2) The superiority of the index over other methods increases with an increasing number of traits under selection, but decreases with increasing differences in relative economic values. Its superiority is greatest when the traits are of equal importance.

(3) The relative efficiency of the index over other methods is much affected by the phenotypic correlations between the traits, being highest when the correlations are low or negative.

Questions other than the relative efficiency in terms of genetic gain come into the choice of method to use. Costs, ease of application and amount of work involved are considerations and will often affect the decision as to which method to use.

Construction of a Selection Index

In Chapter 6, the objectives of improvement are discussed. These can be summarised in the form of an *aggregate breeding value* which combines the traits to be improved, each weighted by its relative economic value. Thus, for a situation where improvement in weaning weight (with relative economic value of a_1) and hogget fleece weight (a_2) is required, the aggregate breeding value of an individual is defined as:

$$G = a_1 \, g_{WWT} + a_2 \, g_{HFW}$$

where the g's are the breeding values of the animal for the individual traits. The problem then is to construct a selection index of the form

$$I = b_1 (WWT - \overline{WWT}) + b_2 (HFW - \overline{HFW})$$

so that the accuracy of predicting the aggregate breeding value is as great as possible; i.e., to determine the values of b_1 and b_2 (weighting factors) so that the correlation between G and I is as large as possible. The items of information required to do this are:

(1) The relative economic values.
(2) The variance of each trait.
(3) The heritability of each trait.
(4) The phenotypic correlation between each pair of traits.
(5) The genetic correlation between each pair of traits.

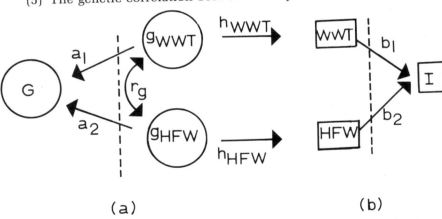

(a)   (b)

Fig. 4.3: Pathways relating aggregate breeding value and the selection index.

Fig. 4.3 shows the relationships between the aggregate breeding value, the individual traits and the selection index. It should be noted that:

(1) To the left of the dotted line (a), the diagram shows the combination of the breeding values for weaning weight and fleece weight, each being weighted by its relative economic value, to form the aggregate breeding value.

(2) In the middle section is shown the relationship between the breeding values of the two traits and the records of the traits themselves.

(3) To the right of the dotted line (b) the diagram shows the combination of the records for weaning weight and fleece weight, each weighted by the appropriate b to form the index I.

From the information given in Table 4.2 and 4.12, the selection index including weaning weight and hogget fleece weight is:

$$I = 4.3 \, (WWT - \overline{WWT}) + 26.9 \, (HFW - \overline{HFW})$$

Hence a sheep with a weaning weight deviation of +1.5 kg and a fleece weight deviation of +0.6 kg will have an index value of

$$I = (4.3 \times 1.5) + (26.9 \times 0.6) = 22.6.$$

It should be emphasised that, if the relative economic values have been calculated as the increase in net return (in cents) resulting from improving each trait by one kilogram, then the aggregate breeding value is measured in cents. The selection index is then a prediction of this aggregate breeding value in cents.

Indirect Selection

In discussing the prediction of breeding values for a single trait, only the records of that trait itself have been used. For example, in predicting the breeding value of a sheep for weaning weight (WWT), only the weaning weights have been included. Selection based on such predictions is often called *direct selection*.

The discussion of genetic correlations between traits on p. 81 shows that it is possible to estimate the breeding value for weaning weight from a trait genetically correlated with it, say, autumn live weight (ALW). The relationships are shown in Fig. 4.4.

The prediction of g_{WWT} can be made from WWT — this is the *direct* path and its accuracy can be measured by the correlation between WWT and g_{WWT} which is equal to $h_{WWT} = \sqrt{h^2_{WWT}}$. If the heritability of WWT is 0.10, then the accuracy is $\sqrt{0.10} = 0.32$. The breeding value for weaning weight can also be estimated from autumn live weight. Using such predictions leads to *indirect selection*. The accuracy of indirect selection is given by the correlation between g_{WWT} and ALW. This is equal to $r_g \, h_{ALW}$ where r_g is the genetic correlation between the two traits and h_{ALW} is the square root of the heritability of ALW. If r_g equals 0.70 and h^2_{ALW} is 0.22, then the correlation is $0.70 \times \sqrt{.22}$ or 0.33.

In this case, the accuracy of indirect selection is slightly better than

85

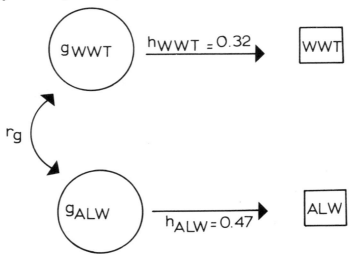

Fig. 4.4: Pathways relating weaning weight and autumn live weight.

direct selection. In general, indirect selection will only equal direct selection in accuracy when $r_g h_{ALW} = h_{WWT}$. Since r_g cannot be greater than one, this will only occur when h^2_{ALW} is greater than h^2_{WWT}. Also generally r_g needs to be high.

Indirect selection is useful if the required trait is very difficult or expensive to measure. For example, clean-scoured fleece weight (which is expensive to measure on the individual sheep) can be improved by selecting on greasy fleece weight. Also in commercial flocks most farmers find it difficult to organise the culling of ewe hoggets on fleece weights at shearing time. Culling on staple length is the next best alternative although only about half as effective at improving fleece weights. Indirect selection is also useful when the trait required can only be measured later in the life of the sheep. This is one reason for investigating the relationships between traits such as number of hogget oestruses and live weight (which can be measured prior to two tooth ewe selection) and number of lambs born (which cannot be normally measured till later).

Even though indirect selection may not be as accurate as direct selection, it may be combined with direct selection to achieve an increase in selection response. For example, the breeding value for WWT can be predicted from both WWT itself and ALW, i.e.,

$$\hat{g}_{WWT} = c_1 (WWT - \overline{WWT}) + c_2 (ALW - \overline{ALW})$$

where c_1 and c_2 are determined as with the selection index so that maximum accuracy in predicting the breeding value for WWT is achieved.

Similarly if the breeding value of ALW is required, then

$$\hat{g}_{ALW} = c_3 (WWT - \overline{WWT}) + c_4 (ALW - \overline{ALW}).$$

It should be noted that the procedure is not limited to just two traits but may include as many characters as are relevant to the selection problem.

The values for the weighting factors (the c's) in the above example are: $c_1 = 0.058$, $c_2 = 0.062$, $c_3 = 0.058$, $c_4 = 0.253$.

Then for a sheep which deviates by +0.2 kg in weaning weight and +0.5 kg in autumn live weight,

$$\hat{g}_{WWT} = .043 \text{ kg and } \hat{g}_{ALW} = .115 \text{ kg}$$

Finally, the use of both the direct and indirect paths for predicting breeding values allows the selection index to be calculated. Thus

$$I = a_1 \hat{g}_{WWT} + a_2 \hat{g}_{ALW}$$

Hence, if $a_1 = 24c$ and $a_2 = 5c$, then the index for the sheep under consideration will be

$$I = (24 \times .043) + (5 \times .115) = 1.61c$$

This procedure has the advantage that the relative economic values are not included until the final calculation and thus can be easily varied to meet the requirements of different breeds and flocks. This is one reason why this method of index calcuation is used in Sheeplan.[7]

Selection on Appearance

Most selection carried out in commercial sheep flocks is on appearance of the animal since no records are available. A considerable amount of selection in ram-breeding flocks is also based on assessing the animal by eye and hand.

No studies have been made in New Zealand to examine the efficiency of the evaluation of sheep in this way. Usually the breeder has a group of characters such as under- or over-shot jaws, twisted feet, black spots and some breed points which are considered faults for which he will cull a sheep despite its merit in other traits. In this, he is operating the method of independent culling levels. Another group of characters includes conformation and wool traits which at present can only be evaluated by eye and hand. Visual assessment of easily measurable traits such as size, fleece weight and staple length are also made. In making his culling decisions on these traits, the breeder is likely to be allowing high merit in some to compensate for deficiency in others and is doubtless also giving due weight to the more economically important traits. Thus he may well be operating a total score approach to selection, a possibility which is suggested by the use of score sheets in some sheep judging.

Experience would indicate that indirect selection based largely on observation of phenotypic associations between traits is also used widely by breeders. For example, many New Zealand breeders believe that there is an association between "quality of bone" and "constitution" and would use an assessment of the former to predict the latter.

87

A further element in selection on appearance is the breeder's idea of *balance* — that the animal is good in most traits but not excessively so in any one of them. This suggests that often the intermediate expression of a trait would be preferred rather than the extreme. In this situation many of the genes affecting the trait will act in an epistatic manner on "balance" even though they act additively on the trait itself. Thus the heritability of "balance" is likely to be low and progress from selection slow.[15]

The major disadvantages of selection on appearance (in situations where objective measurement and records are available) are:

(1) Generally, subjective assessments are likely to have greater variation due to errors of observation than objective measurements. Thus heritability of subjective traits will often be lower.

(2) The individual breeder is likely to vary from time to time not only in his observation of the traits but in the relative emphasis given to each in reaching his culling decision.

(3) There are also differences between individual breeders in their assessments. Hence no general statements can be made about the genetic parameters involved and the effectiveness of the process.

Pedigree Selection

The records of relatives may be used to predict the breeding value of a particular sheep. For present purposes, it is convenient to subdivide the relatives of an individual into three groups:

(1) The direct ancestors of the individual, i.e., its parents, grandparents, great-grandparents, etc.

(2) Collateral relatives such as brothers, sisters, half-brothers and half-sisters, uncles, aunts, etc.

(3) The progeny of the individual. The use of this information will be discussed under progeny testing while the use of the records of the first two groups will be discussed in this and the following section.

A *pedigree* is a record of the direct ancestors of the individual. It is often merely a statement of the tag numbers and breed society registration numbers of the ancestors with no information about their production merit. While such a pedigree can indicate the degree of inbreeding of the animal, it can give little information about its breeding value. Hence in this discussion it is assumed that records on the traits being selected are available on the ancestors, preferably expressed as deviations from the average of their contemporaries.

The use of pedigree information depends on the fact that each individual receives half of its genes from its sire and half from its dam.[14] [15] Even if the genes carried by the sire and dam were known completely, one would still not be able to know the genes received by the offspring because of the chance element at segregation. Because with quantitative traits, it is not possible to know the genotype of the parents, ancestors further back in the pedigree can also assist in predicting the breeding value of the individual sheep.

The amount of attention which should be paid to the records of a particular ancestor depends on:

(1) The *coefficient of relationship* between the ancestor and the individual being assessed. If no inbreeding is involved, a parent has a coefficient of relationship of ½ with its offspring; a grandparent ¼ and a great-grandparent ⅛. Thus generally the influence of an ancestor becomes halved with each additional generation which separates it from the individual being assessed.

(2) The heritability of the trait.

(3) How well the merit of the ancestor is known.

(4) The size of any correlations between the relatives because of similarities in the environment conditions in which they have been maintained.

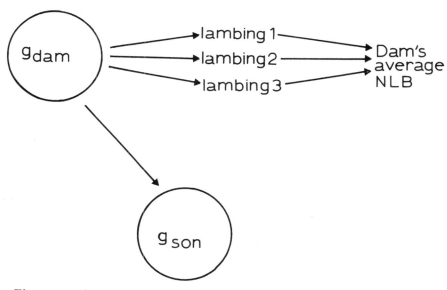

Fig. 4.5: Pathways relating breeding value of a ram to the average of k records of his dam.

A common example in sheep breeding is the use of the average of the records of the dam for number of lambs born to predict the breeding value of her son or daughter (see Fig. 4.5). The prediction equation is:

$$\hat{g}\,(\text{son}) = \frac{½\,kh^2}{1 + (k-1)\,t}(\bar{x} - \bar{\bar{x}})$$

where k is the number of records on the dam, h^2 and t are heritability and repeatability of number of lambs born respectively and $(\bar{x} - \bar{\bar{x}})$ is the average deviation of the dam's records from her contemporaries. For $h^2 = 0.10$ and $t = 0.15$, the values of the regression coefficient relating the deviation to the breeding value are:

k = 1, 0.05; k = 2, 0.09; k = 3, 0.12; k = 4, 0.14; k = 5, 0.16.

Thus for a dam which on three lambings averages 0.2 lamb above her contemporaries in the flock, the breeding value of her son would be estimated as

$$\hat{g} \text{ (son)} = 0.12 \times 0.2 = 0.024 \text{ lamb}$$

It should be noted that the regression coefficient is exactly one-half of the coefficient used in predicting the dam's own breeding value from the average of her records (p. 80). This occurs because the dam passes only a sample half of her genes to her offspring.

The average deviation of the grandam's records could also be used to predict the breeding value of the grandson. Used on their own, the regression coefficient would be one-half of that given above, i.e., for k = 3, it would be 0.06. For records of a great-grandam it would be halved again, i.e., 0.03. This shows why it is seldom worth going back beyond the records of the grandam; the weight given to the information is close to zero and it adds little to the accuracy of predicting breeding values.

It is also possible in predicting the breeding value of a ram to include not only his dam's records but also those of a grandam. If the records are from the paternal grandam, the breeding value of the ram is predicted thus:

$$\hat{g} \text{ (son)} = \frac{\frac{1}{2} kh^2}{1 + (k-1) t} \text{ (Dam's deviation)}$$

$$+ \frac{\frac{1}{4} lh^2}{1 + (l-1) t} \text{ (Grandam's deviation)}$$

where k and *l* are the number of records of the dam and the grandam respectively. However, if the records of the maternal grandam are used, the prediction becomes:

$$\hat{g} \text{ (son)} = \frac{\frac{1}{2} h^2 K(4-h^2 L)}{4h^4 KL} \text{ (Dam's deviation)}$$

$$+ \frac{h^2 L(l-h^2 K)}{4-h^4 KL} \text{ (Grandam's deviation)}$$

$$\text{Where } K = \frac{k}{1 + (k-1) t} \quad \text{and} \quad L = \frac{l}{1 + (l-1) t}$$

The more complicated expression arises because the two sources of information are not independent since some of the information about the maternal grandam is already supplied by the dam herself.

Family Selection

The word "family" is used to denote a group of individuals related to each other.[13] [15] [17] The degree of relationship may vary greatly; e.g., in a full-sib family, the genetic relationship will be ½; in a family of half-sibs, it is ¼. In some breeds, the descendants of some particular male or female which was important in the early history of the breed are called families, e.g., the Duchess family in Shorthorn cattle or the

Erica family in the Angus breed of cattle. In these cases, the genetic relationship is variable and usually very small.

The questions involved in the use of the family average in selection are indicated by the example in Table 4.13, where the fleece weights of three families of half-sisters (daughters of the same sire) are listed. If five hoggets are to be selected on their own record, then the five individuals with fleece weights of 4.6, 4.4, 4.4, 4.2, 4.1 kg (marked by asterisks) would be chosen. If selection were solely on the family average, then the five in Family 1 would be selected. In this case, two hoggets with 4.0 kg fleece weights would be selected. Selection could also be based on the best combination of the individual's own record and its family average. In this case, one of the 4.0 kg individuals in Family 1 may be preferred over the 4.1 kg individual in Family 2, because it comes from a high-performing family. It should be noted that some hoggets will be selected whatever type of selection is used (e.g., the individuals with 4.6 and 4.4 kg in Family 1) because the family average can only be high if a considerable proportion of the individuals in it have good fleece weights.

TABLE 4.13: Hogget Fleece Weights (Kg) of Half-Sister Families
(Daughters of Three Sires)

	Sire 1	Sire 2	Sire 3
	4.0	3.5	4.0
	4.4*	4.1*	3.9
	4.2*	3.6	4.4*
	4.0	3.9	4.0
	4.6*	3.9	3.9
Sum	21.2	19.0	20.2
Average	4.24	3.80	4.04

*Selected on individual merit

In order to use the family average in predicting breeding values, two items of information are needed:
(1) The correlation between breeding values of the members of the family. This is the genetic relationship mentioned earlier and will be designated by R. In sheep breeding, the half-sib family with R = ¼ (in the absence of inbreeding) is the only one of importance.
(2) The correlation between members of the family for the trait being selected. This correlation which will be designated as u will, in the case of the half-sib family, equal $\frac{1}{4}h^2 + C$ where C is a component due to environmental effects which are the same for all members of the family but differ between families.

The prediction of breeding value using family average on its own is then:

$$\hat{g} = \frac{(1 + (n-1) R) h^2}{1 + (n-1) u} (\bar{x}_f - \bar{x})$$

where n is the number in the family average $(\bar{x}_f)$ and $\bar{x}$ is the overall average. If the individual which is being selected is not included in the family average, then the coefficient becomes

$$\frac{nRh^2}{1 + (n-1)\, u}$$

The most likely use of the family average in sheep breeding is in assessing breeding values of rising two-tooth rams for number of lambs born (or reared).[19] In this case, the family information is the average number of lambs born to the n half-sisters of the ram at their two-tooth lambing expressed as a deviation from the average number of lambs born by all of the two-tooths in the particular year. Then assuming that heritability of number of lambs born is 0.10 and that the common environment component is zero, the prediction of breeding value is as follows:

$$\hat{g}_{son} = \frac{n}{n + 39}(\bar{x}_f - \bar{x})$$

since the son has no record to include in the average. When there is no relationship between the sire and dam of the son, the above quantity can simply be added to the prediction of the son's breeding value from the average number of lambs born by the dam. This gives a breeding value based on the combination of the dam's information with the half-sister family average.

Thus for a ram whose dam averages 0.2 lamb above her contemporaries over three lambings and whose 20 half-sisters averaged 1.1 lambs in a year in which all two-tooth ewes averaged 1.0 lambs, the breeding value is estimated as:

$$\hat{g}_{son} = (0.12 \times 0.2) + \frac{20}{20 + 39}(1.1 - 1.0)$$

$$= 0.024 + .034 = +.058 \text{ lamb}$$

The accuracy of family selection will be greatest when u is as small as possible; i.e., when the common environmental component C is zero. Ensuring that all families are grazed together and none is given special treatment will help to keep C low.

The use of the family average is of advantage where (i) heritability of the trait is low, (ii) where the trait is expressed only in one sex as with number of lambs born, and (iii) where the trait can only be measured after slaughter.

Progeny Tests

The principle involved in the use of progeny testing is the use of the average of the *unselected* offspring of an individual to predict its breeding value. In sheep, progeny testing will almost always refer to the testing of sires because dams will seldom have enough offspring to justify its use.

The prediction of the breeding value of the sire is:

$$\hat{g}_{\text{sire}} = \frac{\frac{1}{2}nh^2}{1 + (n-1)\,u} \quad (\bar{x}_s - \bar{\bar{x}})$$

where n is the number of progeny in the average $(\bar{x}_s)$, h^2 is heritability, u is the correlation between the records of the progeny (which are half-sibs), $\bar{\bar{x}}$ is the average of the records of all progeny of the sires used. The correlation u equals $\frac{1}{4}h^2 + C$ as in the case of half-sib family selection.

The accuracy of progeny testing in estimating the breeding value is given by the following expression:

$$\sqrt{\frac{\frac{1}{4}nh^2}{1 + (n-1)\,u}}$$

The way in which this expression changes is indicated in Fig. 4.6. Comparing I with IV shows that generally progeny testing has most advantage in accuracy over individual selection when heritability is low. Notice, too, that five or more offspring need to be included before the average equals a single record on the individual in accuracy.

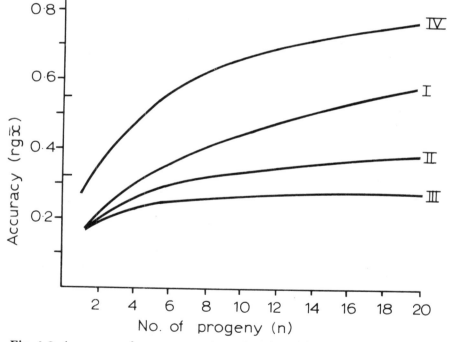

Fig. 4.6: Accuracy of progeny testing when h^2 and C are respectively: I 0.1 and 0; II 0.1 and 0.1; III 0.1 and 0.25; IV 0.3 and 0.

A comparison of I, II and III indicates the effect of the presence of variation caused by environmental effects which are the same for all members of a progeny group but differ from group to group (C). C will have a positive value whenever, either by accident or by design, some sire group is given environmental conditions which are the same for all members of the group but differ from that received by other groups. It can be made small by ensuring that each progeny, irrespective of its sire, receives a random environment. This can usually be achieved by grazing them all together. The need to make C as small as possible is shown by the accuracy of the test. It can be seen that as C increases, the accuracy decreases. Indeed in III where C is 0.25, progeny testing can never be as accurate as a single record on the sire itself.

The use of the progeny average will usually be the most accurate method of estimating the breeding value of a ram. Its effectiveness in producing genetic gain per year is however influenced by the fact that it normally leads to an increase in the generation interval. The reason why this occurs can be suggested by the example in Table 4.14 which compares the age of the sire when his offspring are born under progeny testing and under individual selection.

TABLE 4.14: Number of Progeny (P) at Each Age and its Effect on Generation Interval in Progeny Test and No Progeny Test Situations

Year	Age of sire (A)	No progeny testing P	No progeny testing P x A	Progeny Testing P	Progeny Testing P x A
1970	2 yr	50	100	20	40
1971	3 yr	50	150	—	—
1972	4 yr	50	200	50	200
1973	5 yr	50	250	50	250
Total		200	700	120	490
Generation Interval			$700/200 = 3\frac{1}{2}$ yrs		$490/120 = 4\frac{1}{12}$ yrs

In progeny testing the breeder has to wait till the records on the progeny become available before he can decide whether to keep or cull the ram. Thus, in the example, it is assumed that the ram has 20 offspring born in 1970 and that the trait being selected is hogget fleece weight which will not be measured on the offspring till 12-15 months of age. Hence the ram is first used as a progeny-tested sire as a six-tooth. Because of this delay, he is older when the bulk of his progeny are born and the generation interval is increased. Hence comparison of progeny testing with other selection methods should be on the basis of genetic gain per year, i.e.,

$$\text{genetic gain/year} = \frac{\text{genetic gain/generation}}{\text{generation interval}}$$

Progeny testing will usually result in an increase in the genetic gain/year. However, the generation interval may be increased to such an extent that gain/year is actually decreased compared with individual selection with a shorter generation interval. Mating the ram as a lamb and obtaining the information on the progeny as early as feasible both assist in limiting the increase in generation interval.

Progeny Testing in Practice

In practice, it is not possible to test all rams which are available. Hence initial selection of those to be progeny-tested has to be made on other evidence, e.g., on the dam's average for number of lambs born in dual-purpose breeds or on the individual's own weaning weight (possibly combined with autumn live weight) in meat-sire breeds. Consequently, so-called progeny testing schemes are usually a combination of individual or pedigree selection and progeny testing.

Application of progeny testing usually involves separating the population into two parts: a testing flock and a nucleus flock in which only progeny-tested rams are used. The size of the nucleus flock will affect the number of proven sires needed. Then the problem in the testing flock is to decide how many young sires can be tested. At one extreme, many sires could be tested on few progeny per sire (thus giving intense selection among sires but low accuracy of selection). At the other extreme, few sires could be tested, in which case the accuracy of testing is maximal but there is no room for selection since all sires tested are required. The problem is complex and can only be resolved by a full scale investigation of the variations possible so that genetic progress per year is maximised.[11] [21] This will be considered in detail in the next section.

The two-stage method of selection described above, in which a large number of sires is selected for testing and then a second selection is done on the progeny test, has to be distinguished from a second use of progeny information. In this, no special provision is made for progeny testing. The records on the progeny of each ram used in the flock are summarised as soon as they are available. The information can then be used to select between sires already used in the flock or as the half-sib family averages for use in family selection.

Interpretation of the sire summary (as it is called in Sheeplan[7]) is often unreliable because: (i) particularly with traits measured later in life of the progeny (e.g. later live weight and two-tooth lambing of daughters), the results may be biased by the culling which has taken place among the progeny; (ii) non-random matings of various types may have taken place, so that the average merit of ewes mated to the different sires may be different. Methods of adjusting for the latter problem are available, but the former is difficult to overcome. This emphasises that in a progeny-testing programme in a testing flock, the ewes mated to each sire should be allocated at random and that selection among the progeny should be minimised.

CHOOSING EFFECTIVE SELECTION PLANS

When developing a breeding plan for improvement in a flock or breed, it is helpful to consider the following steps:

(1) Define the objectives of improvement in the flock. This should be done in terms of relative economic values for the traits as in Chapter 6.

(2) Decide on methods of measuring and recording the traits which are to be selected. This is discussed in Chapter 8.

(3) Assemble the information on the variation and heritability of each trait, and the phenotypic and genetic correlations between them. Decide on the procedures to be used in correcting for any major non-genetic differences between sheep in order to increase the accuracy of selection.

(4) Based on the information in (3), choose the selection methods to be used, estimate the amount of progress which is likely to be achieved and, where possible, assess the costs and returns of the competing programmes.

In this section, this last aspect will be considered.

The choice of selection plan depends to a large extent on the amount of genetic gain which can be achieved per year; i.e., on heritability of the trait or index being selected and on the generation interval. In Table 4.15, the accuracy of some of the methods of predicting breeding values have been listed for two traits (one which has the heritability and repeatability of number of lambs born and the other hogget fleece weight). Some general points which help in the choice of procedure are:

(1) If heritability is around 0.30 or above, then selection on the individual's own records is usually sufficient. If repeatability is high (in this case it cannot be less than 0.30), then section 2 of the table shows that there is no real advantage in waiting for later records especially since the generation interval will be increased.

(2) When heritability is lower, then there is need to examine ways of increasing the accuracy of selection. Thus, at heritability of 0.10, a progeny test on 10 progeny increases accuracy by 41% but only by 20% at heritability of 0.30.

(3) Where the trait is expressed in only one sex (as with number of lambs born or reared), where the trait can only be measured after slaughter (such as tenderness of meat), or where the main commercial use of rams is in siring crossbred progeny, progeny testing and use of the half-sib family should be examined. In both (2) and (3), the effect of progeny testing on the generation interval has to be taken into account.

TABLE 4.15: Accuracy of Differing Types of Records in Predicting Breeding Values (Measured by the Correlation Between Records and Breeding Value)

	0.10	0.30
Heritability	0.10	0.30
Repeatability	0.15	0.50
Expressed in:	Female only	Both sexes
1. Single record on the individual	0.32	0.55
2. Average of several records on the individual		
No. of records $\quad$ k = 2	0.42	0.63
$\quad$ k = 3	0.48	0.67
$\quad$ k = 4	0.52	0.69
3. Pedigree records		
(i) Average of k records of dam		
$\quad$ k = 1	0.16	0.27
$\quad$ k = 2	0.21	0.31
$\quad$ k = 3	0.24	0.33
$\quad$ k = 4	0.26	0.34

For grandparents, the figures are halved and for great grandparents divided by 4.

	Female only	Both sexes
(ii) Average of k records of dam and 1 record of the individual		
$\quad$ k = 1	0.35	0.58
$\quad$ k = 3	0.38	0.60
(iii) Average of k records of dam and l records of paternal grandam		
$\quad$ k = 3, l = 3	0.27	0.38
$\quad$ k = 3, l = 4	0.28	0.39
(iv) Accuracy of k records of dam and l records of maternal grandam		
$\quad$ k = 3, l = 3	0.26	0.34
$\quad$ k = 3, l = 4	0.26	0.34
4. Progeny records		
Average of 1 record on each of n progeny		
$\quad$ n = 3	0.27	0.44
$\quad$ n = 5	0.34	0.54
$\quad$ n = 7	0.39	0.60
$\quad$ n = 10	0.45	0.66
$\quad$ n = 15	0.53	0.74
$\quad$ n = 20	0.58	0.79
5. Average of half-sister family		
(i) Average of single records from each of n half-sisters		
$\quad$ n = 10	0.23	0.34
$\quad$ n = 20	0.29	0.39
$\quad$ n = 25	0.31	0.41
(ii) Average of k records of dam and the half-sister average		
$\quad$ k = 3, n = 20	0.38	0.53
$\quad$ k = 3, n = 25	0.40	0.54

Selection for Number of Lambs Born (NLB)

The information in Table 4.15 is used to examine selection plans for NLB, which is the most important trait in the dual-purpose breeds in New Zealand. The heritability of 0.10 and repeatability of 0.15 apply to this trait. It is also a trait which can only be measured in the female. Hence rams can only be selected on the records of female relatives. Rising two-tooth ewes also will not have had a lambing record before entering the flock (unless they lambed as hoggets); so they, too, can only be selected mainly on the records of female relatives.

The possible female relatives available are:

(1) *The dam:* The use of the average of the dam's records for NLB is discussed on p. 89. The accuracy for differing numbers of records is given in section 3 of Table 4.15. On average, the dam will have had 3 lambings at the time of selection of her son; so the accuracy is 0.24. Also the dam is the closest female relative and the use of her records does not alter the generation interval.

(2) *The paternal and maternal grandams:* The sources of information have also been discussed on p. 90. The accuracy is indicated in section 3 (iii) of Table 4.15. Generally, the accuracy is increased by 5-10% by use of the paternal grandam's records and 8% by use of maternal grandam records. It is seldom worth going further back than the grandparents.

(3) *Progeny of a sire:* One would expect the progeny test to be useful in the present case of a sex-limited trait with low heritability. However, it has the effect of greatly increasing the generation interval because of the delay in obtaining the progeny records. For example, a ram born in 1970 if he is used first as a two-tooth will have his first daughters born in 1972 and they in turn will have their first lambing in August-September 1974. Thus the first use of the ram as a progeny-tested sire would be in 1975 when he is 5 years of age. This can be reduced by one year by using the ram first as a lamb. In a typical flock of 3-400 ewes, it is also difficult to utilize a progeny-tested ram adequately; these aspects along with the cost make it difficult to justify except possibly in group-breeding schemes where the utilization of proven sires can be more extensive.[5][20]

(4) *Half-sister family:* For some rising two-tooth rams, lambing records of their half-sisters in the flock will be available.[19] Thus for a sire used first in 1970, the records of his first crop of daughters (as two-tooths in 1972) will be available for the selection of his second crop of sons born in 1971 (but not for his first crop of sons born in 1970). As indicated in section 5 of Table 4.15 the combination of the average of the dam's records with the average of the two-tooth records of the half-sisters gives a useful increase in accuracy (see also p. 93). The main problem in the use of family average is that culling has usually taken place amongst the half-sibs thus causing some bias in the average.

Hence, for ram selection in the average-sized ram-breeding flocks, use of the dam's records combined, where appropriate, with the half-sister family average or the paternal grandam's records is suggested. For ewe selection, the use of the dam's records alone is probably sufficient.

Assessment of Genetic Gain

From pages 73-74 it may be noted that:

$$\text{genetic gain/year } (\triangle G) = \frac{\text{heritability} \times \text{selection differential}}{\text{generation interval}}$$

This expression, in various forms, is the basis for assessing the gain from a selection programme. One convenient form is:

$$\triangle G = \frac{i \, r_{gx} \, \sigma_g}{L}$$

where i is the standardised selection differential read from Table 4.6, σ_g is the standard deviation of breeding values for the trait or index and r_{gx} is the correlation between breeding value and trait or index (the accuracy in Table 15).

The above expression tends to obscure the four ways in which genes, and thus genetic progress, are passed from one generation to the next. The genes of the parents are transmitted to the progeny along the following paths:

	Superiority in breeding value	Generation Interval
Sires to breed sires	I_{SS}	L_{SS}
Sires to breed dams	I_{SD}	L_{SD}
Dams to breed sires	I_{DS}	L_{DS}
Dams to breed dams	I_{DD}	L_{DD}

$$\text{Then } \triangle G = \frac{I_{SS} + I_{SD} + I_{DS} + I_{DD}}{L_{SS} + L_{SD} + L_{DS} + L_{DD}}$$

The superiority in breeding value of each group is given by $i \, r_{gx} \, \sigma_g$. This expression is needed for the analysis of the gains expected in more complicated selection plans, such as progeny testing, where the generation interval, selection differential and the accuracy of selection may differ along each of the above paths.

An example of the use of these formulae is in assessing progeny testing for weaning weight (heritability of 0.10, $\sigma_g = 0.96$ kg).[11][21] Consider a flock of 400 Southdown ewes with 5 age groups requiring 100 2th ewes each year. Eight rams are used; four being 2th and four 4th so that the generation interval L_{SS} and L_{SD} is 2.5 years. L_{DS} and L_{DD} are each 4.0 years. Then assuming an effective lambing percentage of 100%, 100 two-tooth ewes are selected from 200 and i is 0.798. So

$$I_{DD} = 0.798 \times \sqrt{0.10} \times .96 \text{ kg}$$
$$= 0.24 \text{ kg}$$

A 500 ewe Romney flock is available for progeny-testing the Southdown ram lambs. Assuming 20 ewes mated to each ram lamb allows 25 to be tested each year. These are selected on their own weaning weight from the 200 available. The genetic superiority of these will be $1.64 \times \sqrt{.10} \times .96$ kg $= 0.50$ kg (1.64 being i for $25/200 = .125$). As a result of the progeny test, four out of the 25 will be selected for use in the flock as two-tooths. The extra gain achieved by the second selection is complicated to calculate but equals 0.69 kg.

$$\text{Thus } I_{SS} = I_{SD} = 1.19 \text{ kg}$$

$$\text{Then } \triangle G = \frac{1.19 + 1.19 + 0.24 + 0.24}{2.5 + 2.5 + 4.0 + 4.0}$$

$$= 0.22 \text{ kg/year}$$

This is about 40-50% better than selection directly on weaning weight. Much of this superiority arises because in this case, the progeny test information on the ram lambs is available before the normal time of selecting them as two-tooths. Hence there is no increase in the generation interval compared with the normal use of rams first as two-tooths.

As demonstrated, the assessment of genetic gain for a variety of selection programmes and the varying of the flock structure to achieve optimum gains is complicated and is helped greatly by the use of computer simulation.

Economic Appraisal of Selection Plans

Alternative selection programmes may differ markedly in cost, both in initial capital required and in annual running expenditure. Hence in practice, these costs should be taken into account when comparing selection plans.

An important consideration in evaluating a programme is that returns from genetic improvement accumulate over a long period of time and that the pattern of genetic gains (and therefore returns) may be rather variable in the early years. To take account of these aspects, the *discounted cash flow method* has been used. The rationale of the method is indicated by the following. If the current rate of interest is 10%, then $100 invested now at compound interest would become $110 next year, $100 \times (1.10)^2$ in the following year and so on. Conversely $110 obtained next year is equivalent to receiving $100 now or $1 next year is worth $1 \div 1.10 = \$.909$ now and $1 in two years time is worth $\$(.909)^2 = \$.826$ or in 5 years time $\$(.909)^5 = \$.620$ and so on. By the use of this procedure, all returns and expenditures made in different years can be reflected back to the base year and by adding them up, an aggregate profit can be computed up to any year. Thus selection plans which lead to different patterns of returns can be compared.

Little work has been done on this aspect of selection plans in New Zealand. An evaluation of the costs and returns in selecting for fleece production in the Australian Merino has been undertaken.[22]

EXAMPLES OF SELECTION PLANS IN DIFFERENT FLOCKS

Ram-Breeding Flocks of Dual-Purpose Breeds such as the Romney, Perendale and Coopworth

In these breeds, number of lambs born or reared, fleece weight, fleece traits such as colour, unsoundness and cotting, weaning weight as a measure of growth and meat production and spring liveweight are traits which need to be considered in selection. Fleece weight, number of lambs born and reared and live weights are easily measured. Faults in fleeces are usually assessed visually except in special circumstances and, because of variation between observers, are difficult to include in a selection index. Hence, they can best be handled by the method of independent culling levels (p. 81) whereby the breeder sets his own standards for culling on these traits.

Selection methods for number of lambs born (or reared) have already been discussed. Because the heritability of hogget fleece weight is around 0.3 and the repeatability is high, individual selection on this one record is adequate. The heritability of weaning weight in these breeds is also sufficiently high to justify the use of individual selection while generally, the value of later weights such as spring live weight will depend on the extent to which they are genetically correlated with number of lambs, fleece weight and weaning weight. Hence for selection of rising two-tooth rams and ewes a selection index combining the dam's average for number of lambs, with the two-tooths weaning weight and hogget fleece weight is satisfactory. It is usually convenient to estimate the breeding values for each trait using all the information not only on the trait itself but also on the correlated traits (as indicated on p. 86 and as used in Sheeplan).[7] The selection index can then be obtained by summing the breeding values for the traits, each being weighted by its relative economic value.

In using an index for selecting sheep, a sound procedure is to first rank them in order of size of their index. Then inspect the sheep with the highest index for structural soundness and for traits not included in the index which are to be culled by independent culling levels. Then proceed to inspect the second best sheep on index and so on till the required number of sheep is selected. This procedure ensures that the maximum gain will be achieved by the use of the index.

Ram Breeding Flocks of the Meat Breeds

In these breeds, the traits of major economic significance are high growth rate, low lamb mortality in crosses, increased muscling and reduced fat in the carcass. Of these traits, muscling and fatness cannot yet be measured objectively in a simple enough way on the live animal but progress is being made in this field. However, an assessment of the degree of muscling of the animal can be made by skilled observers and this may well be incorporated in a breeding programme. Lamb mortality in crossing is difficult to measure.

In improving growth performance, two cases are distinguished:

(1) where the aim is *early-maturing light-weight lamb production*. In this situation, the measure of growth is the adjusted weaning weight because most lambs would be slaughtered at, or soon after, weaning in commercial practice.

(2) where the aim is *heavy-weight lamb production*. Then the normal time of slaughter would be in the autumn and adjusted autumn live weight is a suitable measure of growth.

The traits already discussed are those that contribute to the returns of the prime lamb producer. There are, however, other traits which will have effects on the costs of production of the rams and will thus affect the breeder's returns. The most important of these is the number of lambs born (or reared) by the ewes in the flock. The amount of importance which this trait is given depends on the extent to which the breeder chooses to reduce his own costs of production as against using all his selection potential to improve the traits which are important to the purchasers of his rams.

As discussed on p. 85, there is a useful increase in accuracy in predicting the breeding value for weaning weight by using not only weaning weight but also autumn live weight. Examples of the estimation of the breeding values are also given on pp. 86-87.

As noted on p. 94, a considerable increase in the rate of gain can be achieved by progeny testing but this will be costly. These costs can be justified better in larger flocks but the flock records of several of the meat breeds indicate that generally flock size is small. Most of the single-entered or stud sires come from a small number of nucleus flocks. Progeny testing, in the larger nucleus flocks is likely to be most efficient in bringing about improvement in the breed as a whole. Another possibility is that of setting up a central testing flock to progeny-test ram lambs contributed by breeders. This would allow the breeders with smaller flocks and resources to contribute to breed improvement.

Selection in Commercial Dual-Purpose Flocks

In a commercial flock, much of the genetic progress will come from the selection of rams. In this selection, there are two important problems. The first is the choice of the ram-breeding flock from which rams are to be purchased. The second is the choice of the rams from within the chosen flock. In theory the two problems may be represented in the following way: The prediction of the breeding value of a ram can be regarded as

$$g_{ram} = b_1 (x - \bar{x}_f) + b_2 (\bar{x}_f - \bar{\bar{x}})$$

where the record of the individual ram is expressed as a deviation from the flock average $\bar{x}_f$ with b_1 being the appropriate weighting factor. The flock average is also expressed as a deviation from the overall breed average $(\bar{\bar{x}})$ and weighted by coefficient b_2. The first part of this equation is equal to the within-flock breeding value calculations which have already been discussed. To calculate the second part, b_2 is required. It depends primarily on the amount of genetic variation between flock averages; information which is not available. However, there are some suggestions which assist in assessing genetic differences between flocks. Generally, where flocks have been part of the traditional hierarchical structure of the breed (see Chapter 7) for a long time, then one would expect between-flock genetic differences to be small (and thus b_2 would be close to zero). On the other hand, where flocks have been selecting consistently and efficiently for increased production using records, then the merit of the flock can be related to the length of time which has elapsed since such selection commenced. The efficiency with which the breeder is applying selection is again difficult to assess. Ideally, one would like to have the selection differentials achieved year after year in selection of the sires in the breeders flock. This is rarely available but some general idea can be obtained by inspecting the records of the sires used in relation to the average of the flock.

Having chosen the flock, selection of the individual rams would follow the pattern already discussed. Because the breeder will be selling rams to a number of clients, the genetic merit of the rams which will be offered to the first client will likely be higher than that offered to later clients. Hence, the choice should be on flocks where selection for performance has been under way for a long period of time and where the buyer can get as near as possible to the first pick of the rams offered for sale.

Selection of rising two-tooth ewes for replacement in commercial flocks will usually be carried out in the absence of records. Exceptions to this could be situations where those ewes which had exhibited oestrus as ewe hoggets have been identified or where provision has been made at hogget shearing to identify those with low fleece weights and unacceptable fleeces. Generally, however, the selection will follow the pattern discussed earlier as selection on appearance. In the main, attention will be given to freedom from structural faults, size, fleece weight as assessed indirectly by staple length and coarseness, acceptable fleece quality and conformation. Culling of older ewes is likely to be for barrenness, lamb deaths, body condition, tooth soundness and disease.

RESULTS OF SELECTION EXPERIMENTS

As indicated in Table 4.4, selection theory results in a prediction of the rate of genetic gain likely to be achieved by selecting for a trait. However, if the theory is to be useful in practice, it is necessary to examine the question: "How well does it predict the size of the response obtained when selection is actually applied to a population?" In the last 25 years a considerable amount of work has been devoted to answering this question.[25]

Measuring the amount of genetic gain achieved by selection is a difficult problem. Taking differences between the averages of successive years or generations in a flock in which selection is being used is unsatisfactory because these averages also contain environmental differences. Thus, changes in a flock average over time cannot prove that genetic progress is taking place. One way of separating out the genetic changes from those caused by the environment is to include along with the selected line, a control group in which no selection is applied. If the control is reared under the same conditions as the selected line, then it is expected that the environmental fluctuations from generation to generation will have the same effect on both lines. Then the difference between the means of the selected and the control line will measure the genetic change or selection response. The control line need not necessarily be unselected; it could be selected in the opposite direction in which case the divergence between the averages measures about twice the genetic change. In some situations, both an unselected- and a downwards-selected control are used to allow assessment of the symmetry of the response in both directions. In addition to these considerations, decisions about the base population from which the selected and control lines are started, the trait or traits to be selected, the selection plan, the number of replications of each line need to be made in setting up a selection experiment.

Most selection experiments have been carried out with laboratory animals such as mice, rats and Drosophila because they are less costly to maintain and have short generation intervals. However, a growing number of selection experiments with sheep are in progress. For New Zealand Romney sheep, the first and most widely known is that established at Ruakura Animal Research Station by Dr L. R. Wallace.[4, 6, 27] In 1948, he formed three flocks, each of 100 mixed-age ewes by selecting sheep from a recorded flock of 1,000 Romney ewes. These consisted of a high-fertility (High) flock chosen for high twinning based on either their own or their dam's records, a control flock selected without any attention to previous lambing performance, and a low-fertility (Low) flock selected against twinning. The flocks which were built up to 130 ewes each, have remained closed and were selected for the same criteria used in their establishment. The selection responses in percentage of lambs born in Fig. 4.7 show a gradual but erratic increase of the High flock over the Control and the Low flock. An analysis of 5 years data (1968-72) showed that the High flock had a greater percentage of lambs born (+ 62%), less barrenness (− 14%), a considerably higher litter size (+ 52% as measured by lambs born/ewes lambing) but a higher level of lamb mortality to weaning (+ 10%) than the Low flock.[6] This resulted in the High flock having a 39% advantage over the Low in percentage of lambs docked. Thus it appears that litter size contributed most to the selection response observed. An assessment of the annual amount of genetic change achieved over the 25 years of selection can be made by comparing the High flock with the Control flock. A difference in docking percentage of 31% indicates an average annual genetic gain of about 1¼ lambs docked per 100 ewes mated.

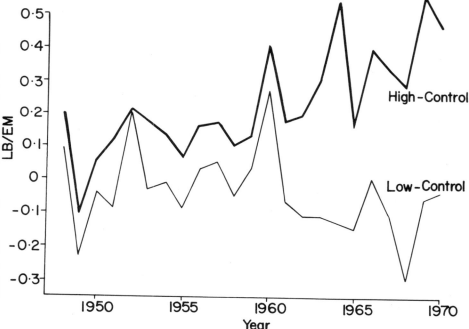

Fig. 4.7: Response to selection for and against twinning at Ruakura[6].

A more detailed examination of the differences between the High and Low flocks was made by outcrossing rams from each flock to commercial Romney ewes and comparing the performance of the daughters so produced.[6] Daughters of the High rams produced 21% more lambs docked than those of Low rams. Because the daughters received only one-half of their genes from the High or Low flocks, this indicates a genetic advantage of 42% which is only little difference from the 39% difference found in the selection experiment. For litter size, daughters of the High rams averaged 26% better, giving an estimate of genetic superiority of 52% which agrees with the result in the selection experiment. However, the outcross daughters showed little difference between High and Low in barrenness and in lamb mortality, suggesting that these two traits are subject to inbreeding depression. Outcross progeny by the High rams produced about 4% less wool than those by Low rams and there was an indication of a small increase in pre-mating body weight in response to the selection for twinning rate. In ewe productivity (combining weight of lamb weaned with ewe fleece weight) there was a 10% advantage for the daughters of High rams over the Low and a 6% advantage over daughters of rams bought from private Romney stud flocks without attention to records. Thus, the experiment clearly demonstrated that selection for twinning rate can lead to worthwhile genetic improvement in reproductive performance. This conclusion is also generally supported by the results of selection for and against the multiple births in Merino ewes in Australia.[25]

A number of selection experiments have also been set up to examine response in wool weights and fleece traits in Merino and Romney sheep.[25] Results from the 9 Merino selection experiments in Australia for higher clean fleece weight show that 8 have resulted in continuing genetic gains. The one exception showed a response to the initial selection of the base animals, but only slight increase thereafter. In experiments where selection has been for higher clean-fleece weight only, the rate of genetic gain has been as high as 7% per annum. Where an upper limit has been placed on fibre diameter, the rate of gain is about 2% per annum and where a lower limit has been placed on crimp number, the gain in clean wool weight has been reduced to 1% per annum, a result which confirms the lower genetic correlation in Merinos between fleece weight and fibre diameter compared with that between fleece weight and crimp number. Other correlated responses to selection for higher clean fleece weight were towards increased staple length and increased yield but little change in body weight and no change in number of lambs born (or weaned). Selection for high and low values for fibre diameter, yield, crimps per inch, and wrinkle score, also showed initial response to the selection of base parents and a continuing response in all cases except the group being selected for long staple length.

Several experiments have been devoted to determining the response to selection for and against body weight at different ages.[16] Results with the Merino confirm the high heritability of body weight at 12-16 months of age. Responses are also obtained in selecting for weaning weight but in this case it is suggested that gains could not be achieved unless selection is based on weights corrected for age of lamb and dam and birth type.

A review of the results of these selection experiments along with many others with laboratory animals suggests the following summary:

(1) During an initial period of about 5-10 generations, the response in the trait being selected, averaged over a number of lines, agrees reasonably well with prediction from knowledge of heritability and selection differentials. Generally, correlated responses in other traits are more variable and are predicted with much lower precision from knowledge of the genetic correlations.

(2) There is often considerable variation in response among different selection lines all derived from the same base population. Especially if the separate lines are small, the variation may result from differing frequencies of the genes in the lines as a result of sampling. Variation in the response of lines from different base populations is caused not only by sampling effects but by real genetic differences between the base populations.

(3) The response to selection is not necessarily the same in both directions. This may be caused by the majority of genes affecting the trait being dominant in one direction, by different selection differentials, by a higher frequency of alleles affecting the trait in one direction or by maternal effects.

(4) The response to selection does not continue indefinitely and the line is said to reach a *selection limit*. If all the genes affecting the trait act in an additive manner, then more and more of the alleles will become homozygous as selection proceeds. The heritability then becomes zero and no further response is expected. In these circumstances, it has been shown that the expected limit depends only on the selection differential, the genetic variance and the effective size of the line (N). The total advance (the difference between the average of the line at the limit, and the initial average) is expected to be 2N times the gain in the first generation of selection, and one half of this total advance will be achieved in 1.4 N generations of selection. For a population with effective size of 20 individuals, it would require about 28 generations (or more than 90 years with sheep) to reach half way to a limit. Thus, even though selection may have been operating on productive traits in the past, it appears that selection limits in sheep arising from this cause will not be of immediate importance.

While the above explanation has been shown to apply in some cases, it is not uncommon to find lines where there is still much genetic variation present although response has ceased. Explanations in these cases involve such possibilities as natural selection opposing the artificial selection being applied to the line, selection favouring the heterozygotes, negative genetic correlations and other complex situations where existing selection theory cannot predict long-term effects.

Thus the results of selection experiments lend considerable support to the conclusion that selection methods as outlined in this chapter are soundly based and can have fruitful application in the sheep industry.

REFERENCES

1 Bowman, J. C. 1974: *An introduction to animal breeding.* Edward Arnold, London, 77 pp.

2 Ch'ang, T. S.; Rae, A. L. 1972: The genetic basis of growth, reproduction and maternal environment in Romney ewes. I Genetic variation in hogget characters and fertility of the ewe. *Aust. J. agric. Res., 21:* 115-129.

3 Chopra, S. C.; Rae, A. L.; Wickham, G. A. 1981: Genotype-environment interactions and genetic parameters in New Zealand Romney sheep. II Heritability estimates at different stocking rates for yearling traits. *N.Z. Jl agric. Res.* (to be submitted).

4 Clarke, J. N. 1972: Current levels of performance in the Ruakura fertility flock of Romney sheep. *Proc. N.Z. Soc. Anim. Prod., 32:* 99-111.

5 Clarke, J. N. 1975: New techniques of reproduction and their relevance to breeding groups. *Sheepfmg A.,* 157-167.

6 Clarke, J. N. 1978: Romneys: highly selected versus industry animals. *Proc. Ruakura Fmrs' Conf.* FPP 160.

7 Clarke, J. N.; Rae, A. L. 1977: Technical aspects of the national sheep recording scheme (Sheeplan). *Proc. N.Z. Soc. Anim. Prod. 37:* 183-197.

8 Dalton, D. C. 1976: *Animal breeding — first principles for farmers.* Aster Books, Wellington, 48 pp.

9 Dalton, D. C.; Rae, A. L. 1978; The New Zealand Romney sheep: a review of productive performance. *Anim. Breed. Abstr., 46:* 657-680.

10 Dolling, C. H. S. 1970: *Breeding Merinos.* Rigby Ltd, Adelaide, 266 pp.

11 Eikje, E. D. 1978: Genetic progress from performance and progeny test selection in Down sheep. *Proc. N.Z. Soc. Anim. Prod., 38:* 161-173.

12 Elliott, K. H.; Rae, A. L.; Wickham, G. A. 1979: Analysis of records of a Perendale flock. II Genetic and phenotypic parameters for immature body weights and yearling fleece characteristics. *N.Z. Jl agric. Res., 22:* 267-272.

13 Falconer, D. S. 1960: *Introduction to quantitative genetics.* Oliver and Boyd, Edinburgh, 365 pp.

14 Lasley, J. F. 1972: *Genetics of livestock improvement.* Prentice Hall Inc., New Jersey, 429 pp.

15 Lush, J. L. 1945: *Animal breeding plans.* Iowa State University Press, Ames, Iowa, 3rd edn, 443 pp.

16 Pattie, W. A. 1965: Selection for weaning weight in Merino sheep. 1. Direct response to selection. 2. Correlated responses in other production traits. *Aust. J. Exp. Agric. Anim. Husb., 5:* 353-360, 361-368

17 Pirchner, F. 1969: *Population genetics in animal breeding.* W. H. Freeman & Co., San Francisco, 274 pp.

18 Rae, A. L. 1956: The genetics of the sheep. *Adv. Genet., 8:* 189-265.

19 Rae, A. L. 1963: Methods of selection for lamb production. *Sheepfmg A.,* 167-181.

20 Rae, A. L. 1974: The development of group breeding schemes: Some theoretical aspects. *Sheepfmg A.,* 121-127.

21 Rae, A. L. 1976: The development of co-operative breeding schemes in New Zealand. In *Sheep Breeding* (Eds G. J. Tomes, D. E. Robertson and R. J. Lightfoot) Western Australian Institute of Technology p. 154-164.

[22] Thatcher, L. P.; Napier, K. M. 1976: Economic evaluation of selecting sheep for wool production. *Anim. Prod., 22:* 261-274.

[23] Turner, H. N. 1969: Genetic improvement of reproduction rate in sheep. *Anim. Breed. Abstr., 37:* 545-563.

[24] Turner, H. N. 1972: Genetic interactions between wool, meat and milk production in sheep. *Anim. Breed. Abstr., 40:* 621-634.

[25] Turner, H. N. 1977: Australian sheep breeding research. *Anim. Breed. Abstr. 45:* 9-31.

[26] Turner, H. N.; Young, S. S. Y. 1969: *Quantitative genetics in sheep breeding.* Macmillan, Melbourne, 322 pp.

[27] Wallace, L. R. 1964: The effect of selection for fertility on lamb and wool production. *Proc. Ruakura Fmrs' Conf.,* 1-12.

[28] Young, S. S. Y. 1961: A further examination of the relative efficiency of three methods of selection for genetic gains under less restricted conditions. *Genet. Res., 2:* 106-121.

5
Mating Plans and their Effects in Sheep Improvement

J. N. Clarke
Ruakura Animal Research Station, Hamilton

SUMMARY

Inbreeding arises in closed flocks of limited size as a result of common ancestry among the parents. In the absence of selection, the rate of inbreeding tends to be proportional to the number of new rams introduced per generation. Under selection it will tend to increase more rapidly, especially as selection becomes more effective or is based on family performance.

To minimise cumulative long-term inbreeding depression and inbreeding effects on selection responses, the size of the population may often need to be increased or the population opened to the judicious introduction of tested rams from outside.

There is little evidence in sheep in support of the deliberate use of inbreeding or linebreeding for enhancement of genetic improvement from selection.

Despite a paucity of information, crossbreeding has played an extremely important role in the development of sheep breeds in New Zealand and continues to prove its usefulness in the production of commercial animals for slaughter. It provides considerable flexibility to sheep breeding programmes permitting rapid changes in the characteristics of a flock.

Some of the limited New Zealand evidence on crossbreeding parameters is reviewed. While operational simplicity favours straight breeding and formation of synthetic breeds, high levels of heterosis for important maternal traits suggest scope for application of rotational crossing.

INTRODUCTION

Breeding plans aimed at changing the genetic make-up of a flock consist of two main parts. The first of these comprises procedures for deciding which animals are to be kept for breeding and provides for predictions of the genetic gains to be expected if the chosen animals are allowed to mate at random. The ways in which chosen animals are actually mated together form the second major aspect of animal breeding plans and are the subject of this chapter.

Random mating implies that an individual of one sex is equally likely to mate with any individual of the other. *Non-random mating systems* are generally broadly classified into four types depending upon genetic or phenotypic relationships among mates:

(1) *Inbreeding* which is the mating of animals which are genetically more closely related than they would be on average under random mating.

(2) *Outbreeding* which involves the mating of animals which are genetically unrelated. Crossbreeding is an extreme example of outbreeding.

(3) *Positive assortative mating* if the animals are phenotypically alike (e.g. mating best to best).

(4) *Negative assortative mating* if they are phenotypically dissimilar (i.e. mating animals of opposite extremes in performance).

INBREEDING

Inbreeding often arises from deliberate and regular full or half-sib, cousin or parent x offspring matings, or from "top-crossing" or "line-breeding" in which a highly considered individual or family is featured regularly in an effort to concentrate its genetic merit in the flock. As a consequence, mates tend to be more closely related to each other than are randomly chosen members of a large population or breeding group. It leads to the same genes being inherited from each of the parents as a consequence of their common ancestry.

Measurement of Inbreeding

The usual criterion of the level of inbreeding is the *inbreeding coefficient*.[44] This coefficient (F), is defined as the probability that the two genes at any loci are identical copies of the same gene inherited from ancestors which the parents have in common. It is thus a measure of the common ancestry of the mates, techniques for measuring inbreeding from pedigree relationships being based on the *degree of common ancestry* observed.[14]

The inbreeding coefficient may range from zero to one (100%). Under random mating in a large population the inbreeding coefficient will remain at the base level of zero. Under non-random mating or restrictions in population size it will increase towards 100% as the number of common ancestors increases to the stage where the mates have all ancestors in common. Formulae are available[14, 42] for estimating levels of inbreeding resulting from regular matings among different kinds of relatives, i.e. where relatives having the same degree of relationship with one another are mated each generation. For example, if offspring are repeatedly backcrossed to one non-inbred parent the coefficients of inbreeding are progressively 0.25, 0.375, 0.438, 0.469, 0.484 in generations 1 through 5, tending towards a maximum value of

0.5. By contrast, for regular full-sib (full brother x full sister) matings they are 0.25, 0.375, 0.500, 0.594, 0.672 respectively tending in this case towards a maximum value of 1.0.

Methods of estimating inbreeding coefficients from pedigree relationships under irregular systems of mating are also available.[14] [42]

Inbreeding and Population Size

Some inbreeding will result, even under random mating, in any breeding group of finite size. Inbreeding "problems" from small populations are recognised and feared by most practising breeders. The inbreeding results from the greater chance of making matings among parents which have ancestors in common as the size of the population decreases.

The *rate of increase per generation* in the average inbreeding coefficient ($\triangle F_g$) due to restrictions in population size is estimated by

$$\triangle F_g \ = \ \frac{1}{2N}$$

where N is the number of individuals which are mated at random in the population.

For different numbers of males (N_m) and females (N_f) the equation becomes

$$\triangle F_g \ = \ \frac{1}{8N_m} + \frac{1}{8N_f}$$

This equation demonstrates that most of the inbreeding comes from the least numerous sex. When, as is common in sheep flocks, there are many more females than males the second term of this expression, $1 / (8N_f)$, has little influence upon the expected rate of inbreeding,

$$\text{i.e.} \ \triangle F_g \ \simeq \ \frac{1}{8N_m}$$

In these equations $\triangle F$ is a proportionate measure of the additional inbreeding relative to the amount of inbreeding necessary before the population would be completely inbred. The equations apply strictly only under random mating and in the absence of selection.

As an example of the effects of population size on the rate of inbreeding to be expected when parents have an equal chance of contributing offspring to the next generation, consider a 400 ewe flock using 2% (.02) of new two-tooth rams each year and having equal numbers of 2, 3, 4 and 5-year old ewes at lambing time. Following the methods presented in Chapter 4, but assuming equal fertility for each age group of ewes, the generation interval on the female side is 3½ years; on the male side it is 2 years. The average generation interval is therefore ½ (3½ + 2) = 2¾ years, i.e. $N_m = \dfrac{2}{100} \times 400 \times 2\tfrac{3}{4} = 22$; for comparison, $N_f = 100 \times 2\tfrac{3}{4} =$ 275 females/generation

Therefore $\triangle F_g \simeq \dfrac{1}{8N_m}$

$$\simeq \dfrac{1}{8 \times 22}$$

$$\simeq 0.0057$$

$$\simeq 0.57\% \text{ per generation.}$$

This is a very slow rate of inbreeding compared with regular systems involving the repeated mating of close relatives.

For sheep, a more useful working formula[42] for calculating the rate of inbreeding *per year* ($\triangle F_a$) is:

$$\triangle F_a \simeq \dfrac{1}{8n_m L^2}$$

where in this case n_m = number of new rams per year

L = average generation interval.

For the above example the average predicted rate of increase in inbreeding ($\triangle F_a$) is

$$\triangle F_a \simeq \dfrac{1}{8 \times 400 \times .02 \times 2.75^2} = 0.21\% \text{ per year (i.e. } 0.57/2.75).$$

Inbreeding in Flocks under Selection

Selection has the effect of increasing the rate of inbreeding because it reduces the effective population size by restricting reproduction to a few animals of similar desired phenotypes. These animals will, to an extent depending on the heritability of the trait under selection, tend to have the same genes. Thus in small populations under selection for a highly heritable trait it will be found that the rams chosen each year will often tend to be the sons of only one or two sires from the previous generation. This will reduce the number of sires contributing to the next generation to below the number that would tend to contribute sons in the absence of selection.

This effect of selection tends to be most marked under family selection (see p. 90) in which animals are selected because they represent some high performing family. The selected animals will have some of the same genes in common and this will increase the proportion of genes which are identical by descent in their progeny.

Formulae for calculating the rates of increase in the inbreeding coefficient for flocks of limited size apply strictly only under conditions of random mating in which all parents have an equal chance of contributing offspring to the next generation. Under continued selection, rates of inbreeding will build up more rapidly than expected from these formulae. The more effective the selection, the greater this discrepancy will be. For individual selection, the effect of selection on the level of inbreeding is proportional to the heritability of the trait and the square of the selection differential.[36] Since the rate of genetic gain from selection is itself proportional to heritability and the selection differential (Chapter 4), methods of maximising genetic gain by increasing the accuracy of selection, increasing the selection differential and reducing the generation interval, also enhance the rate of inbreeding.

Avoidance of Inbreeding

An obvious remedy for inbreeding arising from finite population size is to increase the size of the breeding flock. This will avoid any need to compromise the intensity of selection and may even serve to increase it in very small flocks where maximum reproductive rates of males are not already being achieved (Chapter 4). It will however, serve to increase the size and hence cost of the breeding programme.

Avoidance of the effects of selection in increasing the rate of inbreeding has important experimental application in genetically constant control populations used to monitor selection responses. In these situations the deliberate loss of different sire families can be avoided, and accidental loss minimised, through carefully monitored mating programmes. In addition, the inbreeding arising from variation in family size can also often be minimised by keeping equal numbers of replacement males and females from each family (e.g. one son and equal numbers of daughters per sire family). Under these situations the rise in inbreeding can be reduced by about one third.

Under selection, however, deliberate attempts to reduce the inequality of family size are in direct conflict with the desire to achieve maximum selection intensity. The conflict becomes particularly marked if selection decisions are based upon family information. For example, when selection is based on paternal half-sib performance, maximum selection intensity (through reduction in the number of sire families chosen for breeding) is in direct conflict with full representation of all sire-families in the next generation. Accordingly, in these circumstances avoidance of inbreeding may often require marked increases in population size and, in particular, in the number of rams retained for breeding. The compromise will be less marked when selection is based on individual rather than, or in addition to, family information.

Deliberate attempts are often made to slow down the rate of inbreeding by avoiding matings among close relatives for those animals chosen for breeding. For sheep, avoidance of matings among half-sibs and of sires with their dams or daughters is commonly practised. These systems have the effect of making the inbreeding coefficient more uniform among individuals of any generation but are really only delaying tactics in terms

115

of the rate of inbreeding for the population as a whole.[36] Thus, while the initial rate of inbreeding is reduced, it eventually becomes more rapid in later generations. For livestock populations these tactics are nevertheless sensible; the 'catch-up' phase will often fall beyond the time period of concern to the individual breeder.

The important tactic for minimising inbreeding in small flocks under intense selection is to deliberately avoid the complete closure of the breeding group. Judicious introduction of animals of adequate relative merit from outside the group can reduce the rate of inbreeding without serious consequences to the overall efficiency of selection (Chapter 7). The aim is to increase the *effective size* of the breeding flock without prejudice to the rate of progress from selection. If based upon phenotypic merit such introductions must, however, take account of the genetic merit of the outside flock[22] and of the effects of inbreeding upon phenotypic performance.

Observable Consequences of Inbreeding

Breeders have long recognised the deleterious effects of close inbreeding through the depression of performance. Experiments, particularly with laboratory animals but also with poultry, pigs and other livestock have shown decreases in growth rate, disease resistance, reproductive performance and viability. In general, the more closely the character is related to reproduction and viability (i.e. to fitness in an evolutionary sense) the greater the depression observed from inbreeding. In addition, the depression observed tends to be cumulative, increasing relative to the initial base population as the level of inbreeding is raised.

A review of experiments on the effects of inbreeding on performance in sheep,[42] suggests that ewe progeny of half-sib matings, with $F = 12.5\%$, might have their adult fleece weight lowered as much as 0.25 kg, adult body weight by 3 kg and lambs born by 10 per 100 ewes joined. Their chances of surviving to weaning might be lowered by 20 lambs per 100 born. These results together with those from studies in other mammals have shown that, for maternal traits, the inbreeding decline observed in progeny performance represents a combination of the effects of inbreeding in the progeny and of the effects of inbreeding in the dam on the maternal environment she provides to her young. Both have a depressing effect on the number of lambs born per ewe joined and on the number of lambs weaned per lamb born.[23] Inbreeding of the ewe was more important in the former case and inbreeding of the lamb more important for lamb survival.

The effects of inbreeding on animal performance tend to be in conflict with the effects of selection, particularly in the case of selection for improved reproductive performance because of the pronounced effects of inbreeding on this trait and the need to rely on family or pedigree selection. In effect, the loss in performance from inbreeding must be subtracted from the gain due to selection to measure the net effectiveness of the breeding programme. Studies of the performance of New Zealand Romney sheep from flocks selected for either high or low frequency of multiple births suggested that inbreeding was primarily responsible for the increased pre-weaning lamb mortality of the high-twinning flock.[7] [9] They also suggested that the effects of inbreeding on barrenness may have camouflaged a positively correlated selection response in which selection for twinning had improved the ability of the ram to get ewes pregnant.

Genetic Consequences of Inbreeding

The higher the level of inbreeding the greater the chance that both parents have the same gene descended from a common ancestor. Hence inbreeding increases the proportion of homozygotes and decreases the proportion of heterozygotes. In comparison with random mating this loss in heterozygosity in the population is directly proportional to the coefficient of inbreeding. Thus an inbreeding coefficient of 0.5 predicts that 50% of loci that were heterozygous in the base generation have become homozygous.

This genetic consequence of inbreeding gives rise to the frequently observed increase in frequency of undesirable traits due to recessive genes which were masked in the original population (see Chapter 3). Dominance theory also provides an adequate explanation for the depressed level of expression of traits like growth and wool production observed on inbreeding. Under this theory the depression in perform-ance will be more marked the greater the degree of dominance shown at each locus and the greater the proportion of loci which show dominance in the same direction of increased character expression (*directional dominance*). If there is no directional dominance and only additive gene effects, there will be no reduction in the population mean from inbreeding.

The dominance theory of inbreeding depression is supported by the observation that inbreeding depression tends to be least marked for characters of high heritability. These characters also tend to exhibit a relatively low level of non-additive genetic varation.

Combined Use of Inbreeding and Selection in Genetic Improvement

It is important to realise that inbreeding makes loci homozygous regardless of whether or not these loci are important to breeding objectives. Consequently quite unexpected differences between inbred lines may result from the chance effects of inbreeding on gene frequencies. Thus, characteristics may appear in inbred lines that were quite unknown in the original non-inbred base population. Unfavour-able genes are just as likely to become 'fixed' in a homozygous state as are favourable genes. This is, after all, the basis of the standard practice of mating a sire to his daughters when testing for deleterious recessive genes in sires identified for widespread use through artifical breeding. As a consequence, selection either within or among inbred lines in order to improve their performance or to eliminate those of inferior merit, becomes an important part of any breed improvement programme which aims at exploiting inbreeding.

Early studies on inbreeding and selection in farm livestock were built upon the idea that selection might be capable of offsetting the unfavourable effects of mild inbreeding.[44] Most recent work has shown that artificial selection within inbred lines has been completely ineffective in counteracting the depressing effects of inbreeding on performance.[2] This arises because the random changes in gene frequency under inbreeding increase the proportion of unfavourable alleles made homozygous (i.e. fixed) before selection within the lines can act to reduce their frequencies.[12]

Theoretically, inbreeding offers considerable scope for effective selection among different inbred lines. One of the important genetic effects of inbreeding is that it separates the original non-inbred population into genetically distinct family lines. This is true with respect to both the additive and non-additive genetic variance. Ultimately, under complete inbreeding, animals in each line represent genetic replicates of a single homozygous individual. Thus, at high levels of inbreeding genetic variance shifts from within families to between families and, if large families are formed so that random environmental effects between members of the same family cancel out in their influence on the family mean, selection among them can be very accurate.

In practice, because of the low natural fertility and long generation interval of sheep, the effectiveness of selection among inbred lines is likely to be of limited cost effectiveness, particularly because of the high cost of developing and maintaining inbred lines of poor performance for reproduction and survival. Furthermore, even for more fertile mammals such as pigs, many of the inbred lines have failed to survive more than a few generations of close inbreeding.

The high cost of maintaining even partially inbred lines seriously affects the likely success of a programme of alternate cycles of inbreeding and selection. The aim of these programmes is to use inbreeding to separate out favourable dominance and epistatic effects by selection among the inbred families so produced. The selected families are then crossed and a new cycle of inbreeding started. A major problem, apart from the cost of maintaining the poor-performing inbred lines, is that the whole cycle of inbreeding and selection takes several generations. Thus the benefit of making more effective selection decisions in the one generation on which it is applied, has to be averaged over the whole cycle of inbreeding and selection. Despite these drawbacks, temporary sire-daughter inbreeding and combined family and individual selection in alternate generations could, theoretically, increase annual rates of improvement for growth performance in sheep by the order of 10-15% over those obtainable from mass selection without inbreeding.[12] This prediction has yet to be tested experimentally.

Inbred lines of sheep have been under development at a number of research stations in the U.S.A. It appears from early reports that the main objective of this work is to identify inbred lines that are superior to outbred stock for lamb and wool production when top-crossed to unrelated sheep. Unfortunately no results have yet been published, although most geneticists would now agree that this approach is unlikely to be as cost-effective as the same amount of time and effort devoted to genetic improvement through selection alone.

Breeding programmes involving the deliberate formation and crossing of inbred lines have been followed very successfully for hybrid maize production. The feature of these programmes is that inbred lines are formed and evaluated on the basis of their performance in crosses rather than their performance as inbreds. No new genes are introduced into the population; merely new combinations of existing genes are created. The performance of the inbred line itself is of little importance except insofar as it affects the costs of maintaining the lines. It tends to be of little value as an indicator of crossbred merit. This is to be expected since the method involves a search for lines displaying overdominance in crosses. However, inbreeding levels have to be very high before any useful gains from selection among the crosses can be achieved.

While breeding programmes involving inbreeding, crossing, testing and selection (recurrent selection and reciprocal recurrent selection programmes) have played some role in the commercial improvement of egg production in poultry, they seem to have little future for farm livestock. Even for pigs with their relatively high reproductive rate, little commercial benefit has occurred.[37] This is because of the high costs of systematically testing many crosses in addition to the costs involved in developing and maintaining the inbred lines themselves. Consequently, unless recourse can be made to genetic diversity occurring naturally among existing breeds and strains there would seem to be little prospect for the successful application of these methods in sheep breeding.

Linebreeding

Linebreeding is a form of inbreeding in which animals are mated so that their offspring are kept closely related to some highly regarded ancestor. The aim of linebreeding is to retain the genetic contribution from an admired ancestor at as high a level as possible under the restriction set by Mendelian segregation. A certain amount of linebreeding is inevitable

119

under intense selection based on ancestor performance (e.g. selection on twinning performance of the dam and grand-dam).

In terms of the rate of increase in homozygosity, linebreeding is usually less intense than most regular systems of inbreeding, although this is not always so. Consider, for example, the following mating systems.

```
        Double First Cousins              Half Brother-Sister

   G  H I  J    G  H I  J             G  H I  J
    \ / \ /      \ / \ /
     C    D      E    F            M   N    M   O
     |         |                    \ /      \ /
      \ A     B /                    K        L
         \ X /                        \  Y  /
```

For these pedigrees, $F_x = F_y = \frac{1}{8}$. Thus, both give rise to the same level of inbreeding. The first would usually be classed as inbreeding; the second as "linebreeding to M".[25]

The genetic consequences of linebreeding are the same as those discussed previously under inbreeding: homozygosity is increased and the breed tends to become separated into distinct families each related to some admired ancestor or family. Thus, like inbreeding, linebreeding requires the close association of effective selection with the mating programme if genetic improvement is to be achieved.

ASSORTATIVE MATING

Since likeness in genetic makeup will generally be much less than likeness in phenotype (heritability less than unity), the mating together of animals with similar phenotypes has less effect on the genetic makeup of the offspring than does inbreeding. Further, assortative mating can affect only a fraction of the total number of genes (those that influence the traits considered) while inbreeding and outbreeding can influence the genotypic array at all loci.

Over long periods of time, mating of like to like (positive assortative mating) appears to be a powerful method for creating genetic diversity as it tends to scatter the population towards the two extremes in the character. Any benefits obtained are really attributable to selection which, in practice, will usually discard one extreme. The effects of assortative mating will tend to disappear if random mating is resumed although, in the meantime, the accompanying selection may have brought about some permanent change in gene frequency.

There is little information for farm animals on the benefits of positive assortative mating to selection progress as a conseqeuence of increased genetic variation in comparison with random mating. However, benefits have been observed for laboratory species and some practical sheep improvement schemes have been advocated overseas.[42]

Negative assortative or "corrective" mating on the basis of phenotype has the opposite consequences to positive assortative mating. The population tends to become more uniform for the chosen trait (if heritability is high) than under random mating but returns to normal variability once random mating is resumed; there is a lower resemblance between parents and offspring and between other relatives to a lesser extent; there is little increase in heterozygosity except where variation is due to a few genes with large effects on performance. It may be useful when an intermediate expression of the character is desired, provided one can justify retaining parents of opposite extremes in performance.

OUTBREEDING AND CROSSBREEDING

Outbreeding refers to the mating of animals which are less closely related than the average for the population. The term is a very broad one being used to describe matings between unrelated individuals and between animals of different strains *(strain-crossing)* or breeds *(cross-breeding)*. *Outcrossing* is another term used by breeders to describe the mating of animals less closely related than average. It usually refers to occasional outbred matings followed by a return to a regular system of purebreeding or linebreeding.

In practice, crossbreeding refers to the crossing of individuals from different populations (breeds, strains or lines) which have been isolated from one another for a number of generations. *Backcrossing* is the mating of crossbred animals back to one of the purebred parent breeds used to produce them. *Grading-up* refers to the repeated use for a number of consecutive generations, of sires of one of the original parental purebreds. The aim is to improve the stock represented by the original group of females towards the performance level of the pure breed represented by the sires.

Despite the emotional reactions that have frequently accompanied deliberate departures from purebreeding and traditional concepts of breed purity,[34] crossbreeding has and continues to play an extremely important role in the evolution of sheep breeds (Chapter 2). This is especially true for commercial sheep production.[33] Some of the most highly organised crossbreeding systems occur in the sheep industries of many countries despite a paucity of documented information on the relative merits of alternative breeds and crosses.[4 5 6 41] This is hardly surprising in view of the wide range of functional uses and habitats which characterise sheep production.[33]

Present-day breeds, perpetuated as closed breeding groups, necessarily arise as strains of what were once self-contained genetic groups. They become differentiated genetically in one of several possible ways:[3]

(1) through chance variation in the genes inherited in small populations isolated by geographical, biological or managerial factors (*genetic drift*);

(2) through the forces of *natural selection* leading to gene frequency changes favouring adaptation to prevailing environmental circumstances;

(3) through *artificial selection*, at different intensities or for different objectives;

(4) through gene frequency changes brought about by *immigration* and *crossbreeding*.

Gene frequences are fundamental to the genetic basis of differences between breeds and to the genetic consequences of crossbreeding. Since gene frequencies cannot be observed or even estimated for continuously varying production characters, practical breed descriptions and crossbreeding results consist mainly of production averages, breed type specifications together with information on the variability of these features (Chapter 2).

Since a crossbred inherits half of its genes from each parental breed, its gene frequencies are expected to be intermediate between those of the parental breeds. To the extent that genes have only additive effects the performance of the crossbred will, in the same environment, be intermediate to the performance of the parental breeds. To the extent that genes show non-additive effects the crossbreds will display *hybrid vigour* or *heterosis* and will deviate from the expected intermediate performance level. In the case of a cross between two breeds A and B, heterosis is defined as $\frac{1}{2}(AB+BA)-\frac{1}{2}(A+B)$, where AB and BA represent the performance of first-cross progeny out of B and A purebred females, respectively.

Just as crossbreeding is the opposite of inbreeding, heterosis is the opposite of inbreeding depression and the performance lost on inbreeding tends to be restored on crossing. Heterosis is exhibited in crossbreds simply because existing breeds are, from a genetic point of view, mildly inbred lines. That this is so becomes apparent when one considers the small number of animals upon which introduced breeds have often been founded and the way in which the flocks of a breed tend to be arranged in a hierarchy. Relatively few of the older well-established flocks, which tend to be registered with a breed association and thereby closed to the introduction of genes from outside, supply most of the rams used (Chapter 7). Moreover, this includes only what has happened since breed registration began, which is a comparatively short period on an evolutionary time scale. Thus, while animals within a registered purebred flock may not be very inbred relative to the time when breed registration commenced, straight breeding is nevertheless inbreeding relative to the species as a whole.

Since crossing breeds or strains makes use of naturally occurring genetic diversity and does not involve the cost and effort of deliberately mating and maintaining often poorly adapted inbred lines, it tends to be of greater practical value in the exploitation of heterosis in domestic animals. Furthermore, heterosis in this case represents the improvement that can be obtained over existing, often well-adapted, straight-bred commercial stock, rather than that which can be obtained relative to lowly performing inbred lines.

The Genetic Basis for Heterosis

Just as directional dominance is able to explain inbreeding depression, so too is it an adequate genetic explanation for heterosis.[14] *Directional dominance* requires that the dominant genes at most loci that affect a trait act in the same direction in terms of their effect on performance. For example, at most loci controlling growth rate, genes for faster growth may tend to be dominant. Under this genetic model the amount of heterosis observed in a cross between two populations depends upon the product of the level of dominance and the square of the difference in gene frequency between the populations for each of the loci having the same directional effect on the character in question.

Under the dominance theory crosses between distinct populations, each with individuals homozygous for some recessive genes but differing in the loci for which they are homozygous, would tend to inherit a favourable dominant gene at many loci. Thus undesirable recessives would tend to be masked in the first crossbred (F_1) generation. Just as inbreeding depression tends to be most marked for characters associated with vitality and reproduction, the same is true for heterosis. These characters tend to be modified by constantly changing environmental circumstances throughout the life of the animal. It has been argued that hybrids have greater environmental adaptation through having a larger number of metabolic pathways as a result of their greater number of unlike genes.[16][24] If the young hybrid had even a slight early advantage in the establishment of such physiological adaptive mechanisms as antibody production or temperature regulation, it could conceivably exhibit large heterotic effects at older ages.

Heterosis is expected to be greater if the degree of genetic difference between the parental populations is large. While this tends to be true in crosses between domestic breeds and strains, it often breaks down in wide crosses between populations adapted to markedly different environmental conditions.[14] Overdominance, pleiotropic effects upon several traits, and complex gene interactions involving several loci provide alternative explanations in such cases. Convincing experimental evidence is however lacking since dominance, overdominance, pleiotropic and epistatic effects at individual loci are seldom observable for important performance traits related to adaptation or commercial production.

Although the physiological genetic mechanisms for expression of heterosis are not clearly understood, the directional dominance theory provides a useful working model for summarising the effects on production of crossing different breeds and strains. In the last decade, the theory of breed utilisation through crossing has been much studied on this basis,[11][12][31] and experiments are currently under way to further examine the utility of this approach.

An important aspect of the dominance model of heterosis is the estimate it provides of the performance expected in F_2 (F_1 ram x F_1 ewe) and subsequent generations of crossbreeding. It predicts that the heterosis shown by the F_2 will be only half as great as that shown by the F_1. This decline in F_2 performance to a level half-way between the F_1 and the mid-parent value provides an alternative avenue for estimation of heterosis,[6][11][42] provided that the loss of favourable epistatic combinations can be ignored. In the absence of selection, further change in mean performance is not expected in the F_3 and subsequent generations. Any such change could be indicative of a breakdown in superior epistatic gene combinations present in the parental breeds, as a result of genetic recombination in the gametes produced by crossbred parents.

For important production traits in sheep, the relative amounts of heterosis expected in F_1 and F_2 generations are complicated by the existence of genetically determined maternal effects in addition to the average direct effects of the individual's own genes on performance.[14] Since, the level of heterozygosis in dam and progeny generations is different for the first and later generations of crossbreeding, the level of heterosis in direct and maternal effects is staggered by one generation. The same general pattern of change in heterosis occurs for both the non-maternal (direct) and maternal effects with the latter lagging one generation out of phase.[14] This is illustrated in Fig. 5.1 for a trait exhibiting heterosis of 0.4 and 0.1 units of performance for the direct and maternal effects, respectively.[6] Under this circumstance, assuming the absence of epistatic loss, effects of selection and interactions between direct and maternal effects, the population mean does not stabilise until the F_3 generation is reached. This model also assumes the absence of heterosis for paternal effects through such characters as libido and semen production.

Genetic Parameters of Breed Utilisation in New Zealand

In order to make decisions upon the ways in which breed resources may be utilised most efficiently, it is necessary to have information on the magnitude of the average genetic differences between the breeds, on the importance of hybrid vigour and recombinational loss in crosses derived from them and of the influence of maternal effects for all important aspects of economic performance. These are the basic genetic parameters of breed utilisation, the theory of which has been much extended and rationalised over the past ten years.[11][12][13] Predictions of the outcome of alternative crossing programmes are of very limited accuracy

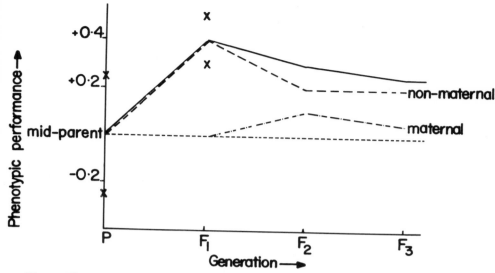

Fig 5.1: Theoretical changes in performance from crossing and interbreeding without selection. The means of the two parent breeds are shown as Xs above P and the means of the two first cross types (which differ due to one breed of ewe being the better mother), are shown above F_1.

unless reliable estimates of genetic parameters are available. Like the parameters which determine responses to selection within a breed, crossbreeding parameters are specific to particular breed and environmental circumstances. Thus, for example, the level of heterosis in a character cannot be predicted in advance for any given cross but can only be assessed by actually making the cross for specific environmental circumstances. Recent studies and experience in various countries have contributed to our general understanding of the relative magnitude of crossbreeding parameters and permit some generalisations to be made.[1] [26] [31] [40] Some of these will be reviewed later in this section. Knowledge of crossbreeding parameters appropriate to New Zealand sheep is very sketchy indeed, despite the important role which crossbreeding has played in the development of our sheep industry (Chapter 2).

Comparison of F_1 Crosses with the Straightbred Parent Population

Comparisons of straightbreds and first-generation reciprocal crosses are being carried out at the Woodlands and Templeton Research Stations. They involve some 1200 pure and first cross ewes derived from the Romney, Border Leicester, Cheviot and Merino breeds at Woodlands and 900 ewes derived from the Romney, Dorset and Corriedale breeds at Templeton. The ewes were generated from 15-25 rams of each breed over the period 1970-73. They are being evaluated for wool and lamb production in carefully balanced matings to meat-breed rams over the years 1972-78. Between 50 and 100 ewes of each pure breed or reciprocal cross are involved. Within each trial all ewes have been run together at all times and are mated to the same rams.

Tables 5.1 and 5.2 present results from the Woodlands trial in 1976,[8] the first year of the trial for which data for adult ewes are available. Ewe reproductive traits have been adjusted for variation in age of ewe. Lamb weaning weight has been adjusted for variation due to sire, sex, age of dam and date of birth, but not for birth/rearing status. The tables show the five main components of lamb meat production as indicated by weight of lamb weaned per ewe (WLW): ewe survival to lambing (EP/EJ), ewe fertility (EL/EP), litter size or prolificacy (LB/EL), lamb survival (LW/LB), and lamb weaning weight. Two indices of overall breeding-ewe productivity for lamb meat and wool (EFW) are also presented. The first of these ($) is based on a weighted average of WLW and EFW, calculated on the assumption that 1 kg of wool is worth four times as much as 1 kg of lamb live weight at weaning; this index is also shown (in parenthesis) relative to the Romney base of 100 in Table 5.1. The second productivity index ($/EBW), expressed relative to purebred Romney ewes at 100, provides a measure of ewe output per unit of ewe body weight prior to the winter. Under the assumptions that a major part of the important variable ewe production costs rests in the provision of winter feed and that winter feed requirements are related to ewe body weight, this ratio would reflect likely performance rankings in terms of productive efficiency.

Table 5.1 presents means for straightbred ewes and means for straightbred plus crossbred ewes averaged over the breed of sire from which they were produced. The sire breed means reflect one half of the average genetic differences among the breeds free from the effects of maternal environment; differences between them indicate the effects on performance of a 50% substitution of genes from one breed of sire relative to another.

The results indicate that Border Leicester ewes were inferior to the other breeds, and to Romneys in particular, in terms of pre-lambing survival and ewe fertility, but more than compensated for this deficiency in terms of prolificacy, lamb survival to weaning and lamb weaning weight. Cheviot ewes were intermediate between Romneys and Border Leicesters for weaner lamb production, falling closer to Border Leicesters as straightbreds than they did on a sire breed basis. In terms of straightbred prolificacy, Romneys were inferior to Cheviots by 0.23 lambs per ewe; on a sire-breed basis the difference was less marked, the difference in prolificacy resulting from a 50% substitution of Cheviot for Romney genes amounting to 0.06 lambs per ewe. The lamb producing inferiority of Merinos over the other breeds arose primarily from low prolificacy although lamb survival was also lower. This inferiortiy of the Merino was, however, less marked in crossbred

TABLE 5.1: Average Ewe Performance — Woodlands 1976

Trait[a]	Straightbred Ewes				All ewes by Sire Breed			
	R	B	C	M	R	B	C	M
Ewes joined (EJ)	64	48	52	46	254	261	232	251
Ewe Survival (EP/EJ)	.98	.92	.97	.98	.99	.97	.99	.99
Ewe fertility (EL/EP)	.97	.88	.96	.93	.95	.97	.96	.97
Prolificacy (LB/EL)	1.25	1.58	1.48	1.05	1.48	1.62	1.52	1.34
Lamb survival (LW/LB)	.83	.94	.84	.80	.88	.89	.89	.86
No. lambs weaned (LW/EJ)	.98	1.19	1.14	.76	1.22	1.35	1.29	1.11
Weaning wt (kg)	22.5	22.6	21.0	20.0	21.8	22.2	21.3	20.9
Wt of lamb weaned (kg)	22.0	26.8	23.9	15.2	26.5	30.1	27.6	23.4
Ewe fleece wt (EFW, kg)	4.17	3.61	2.27	4.00	4.13	3.88	3.21	3.80
Ewe body wt (EBW,kg)[b]	54.7	60.3	53.2	41.2	57.3	59.6	56.9	47.9
Productivity ($)[c]	38.7	41.3	33.0	31.2	43.0	45.1	40.5	38.7
	(100)	(107)	(85)	(81)	(100)	(104)	(94)	(90)
Productivity/body wt[d]	100	97	88	107	100	101	94	107

[a] In this and ensuing Tables the symbols represent: EJ = ewes joined, EP = ewes present at lambing, EL = ewes lambing, LB = lambs born, LW = lambs weaned, R = Romney, B = Border Leicester, C = Cheviot, M = Merino.

[b] Ewe bodywt post-mating

[c] $ = WLW + 4 (EFW)

[d] $/EBW expressed relative to Romney = 100.

ewes than it was in straightbreds. More precise estimates of breed maternal effects will be possible from a comparison of average sire and dam breed effects for all pure and crossbred lambs and will not be attempted until the results from all years of the trial have been collated.

Romneys were superior to the other breeds for ewe fleece weight, the inferiority of Cheviots being very marked although less so on a sire-breed than a straightbred basis. While straightbred Merinos had high fleece weights relative to the other breeds, they were only slightly above average on a sire-breed basis being, in this case, inferior to Border Leicesters.

Allowing for the fact that sire-breed means reflect only half the average genetic differences among breeds, the four breeds tended to exhibit similar relative merit for the overall index of productivity per ewe on a sire-breed as on a straightbred basis. Their relative productivity as straightbreds was: Romney (100), Border Leicester (107), Cheviot (85) and Merino (81). A very different relativity is suggested by the crude "efficiency index" values shown in the final row of Table 5.1. On this basis the difference between Romneys and Border Leicesters is reduced; the Cheviots tended to retain their relative inferiority, while the

Merinos, by virtue of their low body size, were clearly superior to the other breeds.

TABLE 5.2: Comparison of Crossbred and Straightbred Ewes — Woodlands 1976

Trait	Straight-bred (P)	Cross-bred (X)	X/P%	R&B	R&C	R&M	B&M	B&C	C&M
				\<Heterosis in crosses between:\>					
Ewes joined	210	788							
Ewe survival	.96	.99	103	4	2	2	4	2	3
Ewe fertility	.94	.97	104	7	−1	0	7	10	0
Prolificacy	1.34	1.54	115	16	14	18	15	11	18
Lamb survival	.85	.89	105	−1	7	13	3	3	5
Weaning weight	21.5	21.6	100	0	0	−2	3	3	−1
Weight weaned	22.0	28.6	130	28	24	31	37	33	25
Ewe fleece weight	3.51	3.79	108	9	17	0	12	2	11
Ewe body weight	52.4	56.4	108	9	9	5	11	5	6
Productivity	36.1	43.7	121	21	22	17	29	20	19
Productivity/ body weight	100	110	110	10	13	10	16	13	11

The accuracy with which sire-breed means reflect relative breed performance for average direct genetic effects on their progeny depends upon the assumption that heterotic effects on performance are the same for all two-breed crosses. Some idea of the variation in heterotic effects on performance for these crosses is indicated by a further analysis of these Woodlands data presented in Table 5.2. Although the data have yet to be subject to a critical statistical analysis, the level of heterosis displayed by the six pairwise crosses was similar to the average value based on the ratio of crossbred/straightbred performance in the case of pre-lambing ewe survival, prolificacy, lamb weaning weight and ewe body weight. Variation was most marked for ewe fertility, lamb survival and fleece weight, as seen in other published estimates.[31] At Woodlands, Romney-Merino crosses showed below-average heterosis for fertility and weaning weight but high values for prolificacy and lamb survival. Above-average heterosis was shown for ewe fertility by Border-Cheviot crosses and by Romney-Cheviot crosses for wool production. Border-Merino crosses showed above-average heterosis for the composite production traits, weight of lamb weaned and total productivity; Border-Cheviot crosses show high heterosis for lamb production but below-average heterosis for wool production. Very little variation is apparent in heterosis for total productivity if the Border-Merino cross is excluded.

The average levels of heterosis from Table 5.2 are shown in Table 5.3 alongside a recent summary of 23 published reports.[31] In particular, there is a marked discrepancy for prolificacy. This may be partly due to the effects of age. In the earlier years of the Woodlands trial when

TABLE 5.3: Average Heterosis (%) for Performance Traits

Trait	Crossbred vs Straightbred Ewes		F_1 vs F_2 Ewes	Crossbred vs Straightbred Matings
	Woodlands	Nitter[31]	Nitter[31]	Nitter[31]
Ewe Fertility	4	9	12	3
Prolificacy	15	3	18	3
Lamb survival	5	3	15	10
Weaning wt	0	6	9	5
Weight of lamb reared	30	18	51	18
Ewe fleece weight	8	13	21	—
Ewe body weight	8	5	11	—

younger pure and crossbred ewes were compared, the level of heterosis tended to be more marked for ewe fertility and lamb survival but less marked for prolificacy than shown by the above results for the 1976 lambing. Published reports also indicate higher estimates in fertility for younger ewes, probably due to inclusion of heterotic effects on sexual maturity. The high heterosis for prolificacy shown at Woodlands in 1976 and also in other years of this trial are in contrast to the low published results for heterosis in ovulation rate. It does, however, tend to agree with published estimates[31] which are based on a comparison of F_1 and F_2 ewe performance. These are presented in the third column of Table 5.3 which also summarises (column four) published estimates of the heterotic effects exhibited by crossbred lambs as estimated from comparisons of performance in crossbred versus straightbred matings. This summary highlights the superior survival rates generally exhibited by crossbred over purebred lambs.

Comparisons of F_1 and F_2 Generations of Crossbreds

Of particular relevance to the New Zealand scene are the results, summarised in Table 5.4, comparing F_1 and F_2 Border Leicester x Romney crossbreds at Whatawhata from 1959-1967.[17] [18] [19] [20] Compared with the Woodlands results (Table 5.3) Border Leicester x Romney first-crosses at Whatawhata demonstrated less relative superiority over straightbred Romneys for weight of lamb weaned per ewe, but greater relative superiority for fleece weight. For overall productivity, cross-breds were 20-25% higher than Romneys at both locations. At Whatawhata, productivity returned to the Romney level by the F_3 generation following interbreeding in the absence of any artificial selection among ewes.

With the exception of prolificacy, the estimates of heterosis for each of the components of lamb production and for wool production are all higher (by 7 to 10 percentage units) than the average values estimated

TABLE 5.4: Border Leicester × Romney Crossbreeding — Ewe Performance at Whatawhata

Ewe Trait		Romney	F_1	F_2	F_3	Heterosis (%)[a]
Ewes lambing	(EL)	1229	1296	1083	487	
Fertility	(EL/EP)	.88	.93	.88	.86	12
Prolificacy	(LB/EL)	1.17	1.35	1.27	1.28	13
Lamb survival	(LW/LB)	.81	.86	.81	.79	12
Weaning wt (kg)[b]		20.2	21.3	20.4	20.2	10
Wt of lamb weaned (kg)[c]		17.0	22.8	18.6	17.6	58
Ewe fleece wt (kg)		3.49	3.67	3.41	3.41	17
Ewe body wt (kg)[d]		45.4	51.4	48.4	46.8	13
Relative productivity[e]		100	121	104	101	39
Productivity/ body wt		100	107	98	98	21

a   estimated relative to the projected mid-parent value as 200 $(F_1-F_2)/(2F_2-F_1)$
b   per lamb weaned
c   per ewe joined and present at lambing
d   post-tupping
e   calculated as (4 x ewe fleece wt) + weight of lamb weaned per ewe and expressed relative to straight-bred Romneys = 100.

for the 1976 results from Woodlands. This difference is unexpected if, as a number of workers have argued, heterosis estimates based on a comparison of F_1 and straightbreds are biassed upwards as a result of poor adaptation of one or more of the parental straightbred genotypes to the environment in which the animals are tested.[5 28 29 43]

Competitive advantages of crossbreds over straightbreds could be responsible for inflation of heterosis estimates, especially in multi-breed trials in which a high proportion of crossbreds are managed together in the same grazing mob as straightbreds. On the other hand, the difference between the two methods of estimation would be in the direction indicated by the above results if recombination loss in addition to loss of heterozygosity is responsible for deterioration in crossbred performance beyond the F_1 generation. Critical examination of recombination effects has yet to be reported for sheep; the data presented in Table 5.4 are however suggestive of some deterioration in performance from the F_2 to the F_3 generations for all traits except prolificacy and fleece weight.

Since F_1 ewes at Whatawhata were all derived from Romney dams (i.e. the reciprocal cross was not produced), it is conceivable that the F_1 generation at Whatawhata is biased by half the maternal difference between Romney and Border Leicesters. This is unlikely to be a large source of bias for performance traits of mature ewes on the basis of the evidence for maternal effects implicit in the results from Woodlands presented in Table 5.1. However the difference between the F_1 and F_2 means is doubled to provide an estimate of heterosis.[7, 31, 42] The standard error of the estimate is also doubled; sampling variation could, therefore, also account for many apparent differences between the two methods of estimating heterosis.

Crossbreeding Systems for Sheep

There are several comprehensive reviews embracing methods of breed utilisation through crossbreeding and their genetic implications and relevance to sheep.[5] [11] [12] [13] [31] [33]

Grading-up

A 'superior' breed may be expanded in numbers at the expense of another by the continued use of sires of the superior breed over ewes of the inferior breed and then the ½, ¾ and ⅞ etc crosses. This process of grading-up or backcrossing to the superior breed makes use of the reproductive capacity of the breed(s) being displaced. It is therefore relatively efficient in comparison with the alternative approach of lower culling rates within the preferred breed. The rate of progress through grading-up is primarily limited by the number of rams available for the new breed. The process has been used in New Zealand in the past (see Chapter 2).

A consequence of overlapping generations during the early stages of grading-up is an increase in genetic variability because of the mixture of straightbred, halfbred and perhaps three-quarterbred sheep together in the flock at the same time. This is unlikely to be important unless grading-up is being used to bring about a major change in production objectives and there are marked differences in economic returns and management requirements between the two extremes set by the new and displaced breeds.

Regular Crossbreeding Systems

These systems are of several different kinds and offer two important opportunities additional to the exploitation of average genetic differences between breeds. In the first place, they are capable of achieving a balance between individual and maternal effects on performance. This is generally of greatest importance for meat production for which there are two separate phases to the production cycle — the lamb production phase for which puberty, fertility and mothering traits are important, and the growth or carcass production phase of the cycle. In these systems, crossbreeding provides an opportunity to establish a better combination of ewe and lamb traits than can often be achieved within a single breed. This opportunity is often referred to as *complementarity*. The second opportunity offered by regular systems of crossbreeding is the exploitation of heterosis.

Specific Crossbred Combinations: These can make maximum use of differences in genetic merit between the breeds used as male and female parents. However, because sheep have a low reproductive rate in comparison with other domestic species such as pigs and poultry, a relatively high proportion of straightbred matings is required to maintain the parental breed, i.e. sheep have low "excess fertility" available for crossing. This limits the advantages of these systems for sheep.

(1) *Two-breed F₁ crosses*, A ram x B ewe (or AB), make full benefit of heterosis for individual performance and permit a complementation of genes from the breeds used as the male and female parents. A good example of this system in New Zealand is the widely used practice of crossing Down breeds of ram with ewes of the 'white-faced' breeds (Romney, Coopworth, Perendale, Corriedale) for export lamb production (Chapter 2). It involves a two-tier stratification to the sheep industry in which superior average and/or hybrid genetic effects for viability, growth rate and carcass characteristics are introduced from the sire breed. This form of crossbreeding commonly results in improved weaning percentages from increased viability of crossbred lambs or improved lamb production of ewes carrying crossbred young (Table 5.3). Attributes of different sire breeds as crosses for export lamb production are discussed in Chapter 2.

(2) *Three-breed crosses*, T ram x (AB) ewe, of a superior sire breed with the F₁ cross of two dam breeds (A and B) chosen for their average genetic and/or heterotic merit for maternal traits are able to take maximum advantage of hybrid vigour in ewe performance. The terminal sire-breed is usually chosen for its high level of performance in growth and carcass traits, the system thereby also permitting considerable utilisation of heterosis in individual performance.

Three-way crossing systems for meat production are utilised to a large extent in Australia and the United Kingdom, but seldom in New Zealand. In Australia, many slaughter lambs are produced from first cross Border Leicester x Merino (BxM) ewes. First cross BxM ewes are sold to farmers in more reliable lamb producing areas where they are joined to Dorset rams (approx. 5 million annually in N.S.W.). They have a higher level of wool and lamb production to autumn joinings than many other longwool crosses and dual-purpose breeds with which they have been compared,[15] a high level of heterosis for reproductive traits having been demonstrated for this cross in Australia.[28 29 32] In the United Kingdom there occurs a very extensive use of three-way crosses, the Border Leicester breed being used almost entirely for mating surplus ewes of the hardy hill breeds (e.g. Cheviot, Scottish Blackface, Welsh Mountain) to produce a fertile dam for meat production.[1]

(3) *Four-breed F_2 crosses*, (CD) ram x (AB) ewe, also permit maximum utilisation of the average heterosis in individual, paternal and maternal performance and good opportunity to exploit any difference in breeding value for maternal performance between the two breeds used on the female side (A and B) and those used on the male side (C and D). In comparison with three-way crosses there is however more opportunity for recombination loss of superior epistatic gene combinations present in the parental breeds.

Rotational Crossbreeding: This system of mating involves the alternate use, in a regular sequence, of sires of two or more breeds. Since each breed contributes on average to the same extent on the sire and on the dam side of the cross, there is little opportunity for optimum genetic balancing of individual and maternal effects on performance. Consequently selection needs to be for both ewe and lamb performance traits within each of the breeds used and thus there is little opportunity for breed specialisation to take place in comparison with specific crossbreeding. Furthermore, since the genotypes of the progeny are 50% determined by the breed of sire used last, there is less uniformity between generations and hence less opportunity for genetic adaptation to the environment, especially when generations overlap. The method is able to utilise considerable heterosis for both individual and maternal effects, depending upon the number of breeds involved. A two-breed rotation is expected to retain two-thirds of the heterosis for both ewe and lamb traits, a three-breed rotation six-sevenths and a four-breed rotation fourteen-fifteenths of the heterosis. While recombination loss is kept low (the more so the fewer breeds involved) since only female parents are crossbred, high levels of heterosis are the major justification for recurrent cyclical crossing among breeds of similar productive merit. Even in its absence for major component traits, hybrid vigour often may be apparent for overall economic merit.[30]

Combined Specific and Rotational Crossing: In this system a specialised terminal-sire breed is used on ewes produced by rotational crossbreeding (i.e. T ram x AB_{Rot} ewe). It combines some of the advantages of both systems. All the average heterosis in lamb performance is utilised, in addition to a large portion of the heterosis for maternal traits. The system also makes good use of breed differences in maternal and paternal performance, suffers little from recombinational loss and, in common with other rotational systems, requires only sires from each of the parental straightbreds for renewed production of crossbreds.

Formation of New Synthetic Breeds

This system has played a large role in the development of recent sheep breeds. In its simplest form two parental breeds are crossed to produce

133

F_1 progeny which are then interbred (AB_{Syn}) with selection for the desired characteristics in essentially the same way as under straight breeding. In New Zealand this system has been used in developing the Corriedale, Perendale, Coopworth, South Suffolk, South Dorset Down and Borderdale breeds (Chapter 2). For the Polwarth, three-quarterbreds have been interbred. The system has the important operational advantage that it avoids regular recourse to the parental straightbreds and, especially if the number of contributing straight-breds is large, offers some scope for retention of individual and maternal heterosis. As discussed previously, in two-breed synthetics half of the heterosis for both ewe and lamb traits is retained in the F_3 and subsequent generations although the advantage from increased heterozygosity may tend to be squandered through subsequent inbreed-ing if the effective population size is small. The method on its own does not permit the use of separate specialised genotypes for male and female parents. It is also subject to maximum loss in performance from recombination effects, although less so the fewer breeds involved.

A common argument against the interbreeding of crossbreds is that the variation among animals will be increased in the F_2 and later generations over that displayed by the F_1. However, there is usually little noticeable increase in phenotypic variation in most production traits since variation is to a large extent a result of environmental influences.[24] [33] [34] On the contrary, although heterosis might reduce heritability for crossbreds compared with straightbreds,[38] it is often hoped that increased genetic variability in crossbreds might favour the improvement of selection responses or long-term selection limits,[21] [37] in addition to any improvement in selection intensity from higher reproductive rates. There is nevertheless little evidence of improved selection responses in crossbred relative to purebred populations. It would indeed be comforting to believe that selection has been responsible for the superiority observed in Coopworth relative to Romney ewes at Whatawhata,[10] since this amounts to approximately the same level of superiority displayed by F_1 Border Leicester x Romney crossbreds over purebred Romneys in the earlier experiments at this research station (Table 5.4) and is in excess of the predicted level of performance for Border Leicester x Romney crosses interbred without selection!

A variation to the synthetic breed approach is the "gene-pool" concept (see Chapter 7). The aim of this method is to combine a number of breeds, each containing genes desired in the new synthetic, before interbreeding the derived crossbreds. Selection is then applied for the desired traits. The advocates of this approach aim for enhanced selection progress in addition to utilisation of hybrid vigour. Generally the new crossbred flock is retained as a closed breeding group although this is not a necessary requirement. Thus in the United Kingdom the Colbred was produced from the Border Leicester, East Friesian, Clun

Forest and Dorset Horn breeds and the Dam Line developed by the Animal Breeding Research Organisation (breed composition in brackets) from the Finnish Landrace (47%), East Friesian (24%), Border Leicester (17%) and Dorset Horn (12%) breeds.[39] Both these new breeds are aimed at competing with the Border Leicester as a sire to mate with surplus hill ewes to produce a special-purpose dam with improved maternal characteristics for slaughter lamb production.

A further variation on the theme of synthetic breed formation involves the introduction of a particular desired characteristic into an existing breed. Thus the Poll Dorset was developed from the Dorset Horn by crossing with Corriedales or Ryelands to introduce genes for polledness into the breed. This was followed by backcrossing to Dorset Horns combined with continued selection for lack of horns. This method applies particularly to characteristics known to be determined by a small number of genes with easily recognisable effects.

Possible Future Applications of Crossbreeding

It is clear from the earlier review that the cumulative effect of heterosis for each of the component traits contributing to weight of lamb weaned per ewe is considerable. This favours the adoption of regular crossbreeding systems, in comparison with synthetic breed formation and straightbreeding. There are however a number of difficulties to the application of regular crossbreeding systems for sheep.[31]

One of the major difficulties to the exploitation of heterosis is that of ensuring that a high proportion of the slaughter lambs and the ewes which produce them are crossbred relative to the total number of animals within the population. Whereas synthetic and rotational systems require only the one female genotype for the production of both slaughter lambs and ewe replacements, three-way crossbreeding, T x (AxB), requires that the total production unit comprises two types of ewes, F_1 crossbreds *and* straightbred parental flocks of breed B for the production of F_1 ewes, in addition to the ewes required to generate straightbred A and T sires. As a consequence, within the total production unit, the degree of increased heterozygosity over purebreeding is limited by the purebred matings (AxA, BxB and TxT) and two-way matings (AxB) necessary to generate crossbred ewes and lambs. Its efficiency in utilising heterosis is therefore dependent upon ewe reproductive rate, the length of reproductive life and the culling rate among young replacement females. These characteristics determine the excess reproductive potential available for crossbred ewe production. For typical New Zealand circumstances (one lamb weaned per ewe exposed, a reproductive life of four years and a culling rate of one-third among ewe replacements), only 50% of all slaughter lambs and 28% of all ewes can be crossbreds. If, however, the reproductive life is five years and only one-quarter of the replacements are culled, the proportions increase to 86% and 54%, respectively. If at the same time the reproductive rate is 1.8 lambs per ewe, the proportions become 96% and 75%, respectively.[31]

Because of the importance of ewe feeding and management costs to total non-fixed sheep farming expenses, the economic efficiency of lamb meat production tends to be inversely proportional to the number of breeding ewes required to produce a given weight of lamb carcass. Theoretical studies of the relative efficiencies of different mating systems may, therefore, be usefully undertaken on this basis.[31] Results appropriate to typical New Zealand circumstances are summarised relative to straightbreeding in Table 5.5. They have been evaluated either with or without a 20% growth superiority from a terminal sire breed, assuming average values for individual and maternal heterosis obtained from an extensive review of published literature but zero average genetic differences between alternative straightbred genotypes.[31]

TABLE 5.5: Relative Efficiencies of Crossbreeding Systems for Lamb Meat Production in Relation to Growth Rate of the Terminal Sire: (Nitter[31])

| Mating System | Growth Superiority of Terminal Sire Breed[1] | |
	0	20%
Straightbreeding	1.00	1.00
Two-breed synthetic	1.24	1.24
Three-breed synthetic	1.32	1.32
Two-breed rotation	1.32	1.32
Three-breed rotation	1.43	1.43
Terminal crossing to:		
Straightbred dam	1.07	1.11
F_1 dam (three-way cross)	1.19	1.24
Two-breed synthetic	1.29	1.35
Two-breed rotation	1.37	1.43

[1] Over the maternal breed or crossbred

As expected, all crossbreeding systems are clearly superior to straight-breeding, especially those capable of exploiting heterosis in maternal performance in addition to heterosis in lamb traits. Thus, a terminal cross to a straightbred dam offered the least of all mating systems studied, although its superiority relative to straightbreeding was increased the higher the average genetic growth superiority of the terminal sire breed and the higher the lamb production and longevity of the dam. Similarly a two-breed synthetic offered more than a two-breed terminal cross by virtue of the opportunity it provides to exploit heterosis for maternal performance.

The three-way cross showed no advantage over a two-breed synthetic even when the average growth superiority of a terminal sire-breed is as high as 20%. However, if the average reproductive performance of the

136

straightbreds was as high as 1.4 lambs weaned per ewe, the three-way cross became superior to the two-breed synthetic, its advantage increasing as the length of the reproductive life of the ewes and the growth superiority of the terminal sire-line were increased.

Only under high excess reproductive potential was the three-way cross system competitive with a three-breed synthetic or a two-breed rotation. The latter two systems had the same relative efficiencies at all combinations of circumstances investigated, their efficiency relative to purebreeding (approximately +30%) changing very little with weaning percentage, length of reproductive life and replacement culling rates. Only under very high reproductive rates was the three-way cross competitive with a three-breed rotation or a terminal rotation system.

The three-breed rotation was always superior to a system involving terminal crossing to a two-breed (dam breed) rotation unless there was an advantage to be had from average growth superiority of a terminal sire breed. In this situation and low excess reproductive potential, the latter mating system (TxAB$_{Rot}$) always showed considerable superiority over all others.

The assumptions underlying this theoretical study must be born in mind.

(1) Wool production and the efficiency with which different breeds and crosses gather and utilise forages for lamb and wool production have not been considered.

(2) The levels of hybrid vigour assumed are average values but there is a marked variation in estimates especially with regard to maternal heterosis for fertility and heterosis for wool production.[31]

(3) Possible recombination losses have not been taken into account. They would reduce the relative efficiency of synthetics and, to a lesser extent, of rotational systems.

(4) The possibility of average genetic differences among dam breeds has not been allowed for. These could permit an integrated matching of breed characteristics with variation in environmental features of climate, nutrition, management and product markets within the overall production system as illustrated by the breed stratification in the British sheep industry.

Operational Considerations

Tiered industry breeding structures may often suffer from a lack of vertical integration between the tiers. In the case of three-way cross (TxAB) there needs to be feed-back of incentives for genetic improvement from the commercial environments in which the crossbred ewes are run, to the flocks in which they are produced and, even further back, to the flocks in which the A and T rams and B ewes are bred. The greatest scope for selective improvement in the reproductive performance of the crossbred ewe rests in the flocks supplying rams of breed A. Incentives are needed within the system to ensure that breeders of AxB ewes obtain breed A rams of a high breeding merit for reproductive traits. These traits will only be expressed some years later and will have greatest immediate benefit to the slaughter lamb producer. These industry problems are discussed further in Chapter 7, selection and crossbreeding being complementary rather than alternative tools for genetic improvement of animal production.

Other non-genetic considerations are also relevant to the industry-wide efficiency of any stratified crossbreeding system. For example, allowance must be made for possible variation in the ram producing costs for different breeds in different environments, while organisation-al difficulties of managing different separate mating groups within a single farm also impose extra costs. Thus, rotational crossbreeding, while having considerable theoretical advantages, poses particular organisational problems, especially at tupping. Because of the existence of overlapping generations within the breeding flock, several breeds of ram are required in each year. Thus several separate mating groups need to be formed. In addition ewes require to be of identifiable breed composition requiring the marking of lambs according to their breed genotype between lambing and weaning.

TABLE 5.6: Relative Genetic Contributions Under Two-Breed Rotational Crossing (AXB) Commencing with Ewes of Breed B

Year	Proportion of Breed A Rams Required	Proportion of Genes from Breed B	
		Mean	Standard Deviation
1	1.00	1.00	0
2	1.00	1.00	0
3	.75	.88	.22
4	.50	.75	.22
5	.31	.64	.22
6	.19	.55	.10
7	.36	.59	.12
8	.56	.63	.14
9	.66	.65	.15
10	.64	.62	.17
11	.56	.58	.18
12	.44	.53	.23
13	.39	.51	.17
14	.42	.51	.16
15	.49	.53	.16

Table 5.6 presents the proportion of breed A rams required during the first 15 years of two-breed rotational crossbreeding initiated from ewes of breed B. The remaining rams are, of course, all of breed B. The figures apply to a simplified set of circumstances in which a flock is

assumed to comprise equal numbers of 2, 3, 4 and 5-year-old ewes. Also shown in Table 5.6 is the proportion of genes from breed B represented in the flock each year. This fluctuates considerably from year to year, especially in the early years. The between-animal standard deviation in the proportion of genes from breed B also fluctuates from year to year. It will be responsible for variation in average genetic merit within the flock each year. For this reason, rotational crossing is likely to capitalise on the benefits of increased heterozygosity only if there exist negligible average genetic differences for overall economic performance between the breeds used. Variation in a single performance trait (e.g. wool quality) will not matter provided that any increase in production costs, or decrease in returns, is counterbalanced by variation in some other component of financial return.

The variation observed in a single performance trait may in fact allow rotational crossbreeding to be applied quite simply. For example it might be feasible to obtain a sufficiently accurate identification of breed genotype for individual animals from an easily recognisable wool or body conformation trait which exhibits a large additive difference between each of the two (or more) breeds used in the rotation. In such a situation animals could be easily drafted into their appropriate mating group on the basis of their visual appearance and the need for breed pedigree records be eliminated. Traditional beliefs in the merits of visual uniformity are likely to discourage any widespread commercial adoption of this possible approach.

Other non-systematic rotational crossing procedures have also been proposed.[40] On the basis of the large superiority of crossbreds over straightbreds observed for lamb production in North America (12% for two-breed crosses, each additional breed giving an additional 8-20% gain up to the four-breed cross), a system based on the commercial use of crossbred rams has been advocated. The aim is that straightbred ewes would be used only to produce crossbred rams. The two-breed cross ewe lambs would be either sent for slaughter or kept for replacements to generate three-and four-breed crosses. It is envisaged that a decision would be made each time there arose a need for additional crossbred rams, on the best breed constitution likely to match the average breed constitution of the ewes on hand at that time. In this way a high proportion of crossbred ewes would be maintained in the sheep population. It is considered that up to the stage of a four-breed cross the gains from heterosis would tend to more than counterbalance any loss from inclusion of breeds of low average genetic merit and that highly prolific breeds like the Finnish Landrace could be used to help maintain a balance between meat and lamb producing abilities in the offspring.

CONCLUSIONS

It is very evident that the key feature of crossbreeding is the large degree of flexibility it provides to meet changing economic and market circumstances. Crossbreeding is the only way in which the latent flexibility offered by breed resources can be utilised rapidly. There would seem little doubt that in the future the application of crossbreeding will lead to a very much higher level of specialisation in the sheep industry of this country, with crosses being "tailor-made" to meet specific requirements.

If the full flexibility provided by crossbreeding is to be retained for the future, it is important that efforts be made to conserve those purebreds having desirable characteristics. This becomes a costly venture because of the difficulty of recognising genes which, while not immediately competitive, may become useful as market and other environmental circumstances change. In New Zealand there is the good example of the Drysdale breed, derived from an undesirable strain of the New Zealand Romney, now fulfilling an important role in our sheep industry. There would seem to be large scope for international collaboration to this problem of breed conservation.[27] [35]

This review has demonstrated the many ways in which breeds might be utilised through crossbreeding. In view of the paucity of documented information in New Zealand it is clear that the subject is a fruitful area for further research. It is to be hoped that commercial producers will themselves try alternative systems on a practical scale, that their efforts will be guided by the promising leads reviewed above and that they will be coupled with parallel research efforts to document responses in a way that will provide information which can be interpreted in terms of the basic parameters of breed utilisation. Only in this way will the results be capable of meaningful extrapolation to other crossbreeding systems and circumstances. Unfortunately this has not been a feature of past research, many trials being poorly designed, controlled and documented. It should be a mandatory requirement for all future research, since testing even the most promising of all possible crosses and crossbreeding systems is a major undertaking, there being many possible combinations and large numbers of animals are required. The confounding of selection responses with crossbreeding effects is a further major difficulty, while the question of selection responses in crossbred relative to straightbred populations is itself an area where research information is sadly lacking.

Acknowledgements

To Drs A. H. Carter and G. A. Wickham for helpful suggestions in the preparation of this paper and to colleagues and staff of the Ruakura Genetics Section collaborating in the design (A. H. Carter), conduct (J. L. Dobbie, A. E. Uljee) and analysis (S. M. Hickey, H. A. Templer) of the Woodlands crossbreeding experiment.

REFERENCES

[1] Bichard, M. 1974: Crossbreeding and utilisation of breed differences in sheep production — review of the present situation in Europe. *Proceedings 1st World Congress of Genetics Applied to Livestock Production.* 1: 779-784

[2] Bowman, J. C. 1974: *An Introduction to Animal Breeding.* Studies in Biology No. 46, Edward Arnold Ltd, London.

[3] Carter, A. H. 1972: New blood sought for N.Z. sheep flocks. *N.Z. Jl Agric.* 125: 25-53.

[4] Carter, A. H. 1975: Importation and utilisation of exotic livestock breeds. *Proc. 3rd World Conf. Anim. Prod., Melbourne,* 1973. (Ed. R. L. Reid) Sydney Univ. Press: 608-615.

[5] Carter, A. H. 1976: Exploitation of exotic genotypes. In *Sheep Breeding.* (Eds G. J. Tomes, D. E. Robertson, and R. J. Lightfoot) Western Australian Institute of Technology, Perth: 117-128.

[6] Clarke, J. N. 1971: Crossing experiments for the utilisation of sheep breeds in New Zealand. *Proc. N.Z. Soc. Anim. Prod.* 31: 164-74.

[7] Clarke J. N. 1972: Current levels of performance in the Ruakura fertility flock of Romney sheep. *Proc. N.Z. Soc. Anim. Prod.* 32: 99-111.

[8] Clarke, J. N. 1977: In *Annual Report of the Research Division 1976-77:* Ministry of Agriculture and Fisheries, Wellington: 228-230.

[9] Clarke, J. N. 1978: Romney sheep — highly selected versus industry animals. *Proc. Ruakura Fmrs. Conf.:* Aglink FPP160.

[10] Dalton, D. C.; Clarke, J. N.; Rattray, P.V.; Kelly, R. W.; Joyce, J. P. 1978: Sheep breeds — comparison of Romneys, Coopworths and Perendales. *Proc. Ruakura Fmrs. Conf.:* Aglink FPP132.

[11] Dickerson, G. E. 1969: Experimental approaches in utilising breed resources. *Anim. Breed. Abstr.* 37: 191-202.

[12] Dickerson, G. E. 1973: Inbreeding and heterosis in animals. In *Proceedings of the Animal Breeding and Genetics Symposium in Honour of Dr Jay L. Lush,* American Society of Animal Science Champaign, Ill. U.S.A: 54-77.

[13] Dickerson, G. E. 1974: Breed evaluation. In *Proceedings Working Symposium on Breed Evaluation and Crossing Experiments with Farm Animals,* Zeist, Netherlands: 7-24.

[14] Falconer, D. S. 1960: *Introduction to Quantitative Genetics.* Oliver and Boyd, Edinburgh, U.K.

[15] Fogarty, N. 1978: Development of the ideal prime lamb dam. *Wool Tech. and Sheep Breed.* 25: 31-36.

[16] Hazel, L. N. 1961: Extent and usefulness of hybrid vigour in animal improvement. In *Germ Plasm Resources,* (Ed. R. E. Hodgson) Publication No. 66. American Association for the Advancement of Science, Washington, D. C.,

[17] Hight, G. K.; Jury, K. E. 1970: Hill country sheep production. I. The influence of age, flock and year on some components of reproduction rate in Romney and Border Leicester x Romney ewes. *N.Z. Jl Agric. Res.* 13: 641-659.

[18] Hight, G. K.; Jury, K. E. 1971: Hill country sheep production. III. Sources of variation in Romney and Border Leicester x Romney lambs and hoggets. *N.Z. Jl Agric. Res.* 14: 669-686.

[19] Hight, G. K.; Jury, K. E. 1973: Hill country sheep production. IV. Ewe live weights and the relationship of live weight and fertility in Romney and Border Leicester x Romney ewes. *N.Z. Jl Agric. Res.* 16: 447-456.

[20] Hight, G. K.; Atkinson, J.J.; Sumner, R. M. W.; Jury, K. E. 1976: Hill country sheep production. VII. Wool traits of Romney and Border Leicester x Romney ewes. *N.Z. Jl Agric. Res.* 19: 197-210.

141

21 Hill, W. G. 1971: Theoretical aspects of crossbreeding. *Ann. Genet. Sel. Anim.* 3: 23-34.

22 James, J. W. 1977: The theory behind breeding schemes. In *Sheep Breeding* (Eds G. J. Tomes, D. E. Robertson and R. J. Lightfoot) Western Australian Institute of Technology, Perth: 145-153.

23 Lax, J.; Brown, G. H. 1968: The influence of maternal handicap, inbreeding and ewe's body weight at 15-16 months of age on reproduction rate in Australian Merinos. *Aust. J. Agric. Res.* 19: 433-42.

24 Lerner, I. M. 1954: *Genetic Homeostasis.* Oliver and Boyd Ltd, Edinburgh.

25 Lush, J. L. 1943: *Animal Breeding Plans* 2nd Edn, Iowa State University Press, Ames.

26 Maijala, K. 1974: Breed evaluation and crossbreeding in sheep — a summarising report. *Proceedings of the Working Symposium on Breed Evaluation and Crossing Experiments, Zeist, Netherlands,* 1974, 4:-389-405.

27 Mason, I. L. 1974: The conservation of animal genetic resources. *Proc. 1st World Congr. On Genetics Applied to Livestock Prod.,* Madrid 2: 13-19.

28 McGuirk, B. J. 1967: Breeding for lamb production. *Wool Tech. Sheep Breeding.* 14: 73-75.

29 McGuirk, B. J. 1977: Heterosis in crosses between sheep breeds. *Proceedings 3rd International Congress of the Society for Advancement of Breeding Research in Asia and Oceania (SABRAO),* Canberra, 6: 19-22.

30 Moav, R. 1966: Specialised sire and dam lines. I. Economic evaluation of crossbreds. *Anim. Prod.* 8: 193-202.

31 Nitter, G. 1978: Breed utilisation for meat production in sheep. *Anim. Breed. Abstr.* 46: 131-143.

32 Pattie, W. A.; Smith, M. D. 1964: A comparison of the production of F_1 and F_2 Border Leicester x Merino ewes. *Aust. J. Exp. Agric. and Anim. Husb.* 4: 80-85.

33 Rae, A. L. 1952: Crossbreeding of sheep. *Anim. Breed. Abstr.* 20: 197-207, 287-299.

34 Rae, A. L.; Wickham G. A. 1970: Crossbreeding and its part in the utilisation of existing and introduced breeds. *Sheepfmg A.* 87-100.

35 Rendel, J. 1975: The utilisation and conservation of the world's animal genetic resources. *Agric. and Environ.* 2: 101-119.

36 Robertson, A. 1961: Inbreeding in artificial selection programmes. *Genet. Res.* 2: 189-194.

37 Robertson, A. 1971: Optimum utilisation of genetic material with special reference to crossbreeding in relation to other methods of genetic improvement. *Rep. 10th Int. Congr. on Anim. Prod., Paris-Versailles:* 57-68.

38 Salah, E., Galal, E.; Hazel, L. N.; Sidwell, G. M.; Terrill, C. E. 1970: Correlation between purebred and crossbred half-sibs in sheep. *J. Anim. Sci.* 30: 475-480.

39 Smith, C. 1978: Prolific line of sheep. *N.Z. Jl Agric.* 137: 25-26.

40 Terrill, C. E. 1974: Review and application of research on crossbreeding of sheep in North America. *Proceedings 1st World Congress on Genetics Applied to Livestock Production, Madrid* 1: 765-777.

41 Turner, H. N. 1967: Genetic aspects of characteristics to be considered in the evaluation and utilisation of indigenous livestock. *World Rev. Anim. Prod.* 3: 17.

42 Turner, H. N.; Young, S.S.Y. 1969: *Quantitative Genetics in Sheep Breeding.* Macmillan, Australia.

43 Wiener, G.; Hayter, S. 1975: Maternal performance in sheep as affected by breed, crossbreeding and other factors. *Anim. Prod.* 20: 19-30.

44 Wright, S. 1921: Systems of mating. *Genetics* 6: 109.

6

Objectives for Sheep Improvement

C. A. Morris[1], J. N. Clarke[1] and K. H. Elliott[2]
[1]Ruakura Agricultural Research Centre, Hamilton
[2]Wool Research Organisation of New Zealand, Christchurch
Consultants: B. A. Bell and D. L. Johnson

SUMMARY

Objectives for sheep improvement have been defined for four major classes of breeds: dual purpose breeds of three types (coarse-woolled breeds, fine-woolled breeds and specialty carpet-woolled breeds) and prime lamb sire breeds. Only those objectives which are expected to respond to genetic selection are described. Emphasis has been placed on those aspects of production which will change most in terms of their contribution to flock net income. For the prime lamb sire breeds, these aspects are lamb live weight at slaughter, and lamb survival in crossbred progeny from birth to slaughter. For the three classes of dual purpose breeds the most important traits to improve are lambs weaned per ewe mated, fleece weight and weaning weight, in that order. In addition, staple length and fibre diameter are important for fine-woolled breeds. Medullation, bulk, lack of lustre and whiteness have a place in specialty carpet-woolled breeds, but it is not known how they should be placed in order after fertility and fleece weight. Structural soundness in all breeds is stressed, although it is not known how important it is, nor how its various aspects can be measured reliably under experimental or field conditions.

INTRODUCTION

The *national objective* for sheep improvement is to increase export earnings. The *farmer's objective* is to derive extra satisfaction and income from sheep production. As satisfaction is difficult to quantify, a simpler criterion of improvement is an increase in net farm income. On farms which do not change in effective area from one year to the next, this criterion is identical in practice to increasing net income per hectare.

This chapter will concentrate on sheep improvement in relation to the commercial farmer's objective of increasing net income per hectare. There are many aspects of production which contribute to net income

143

per hectare. Managerial decisions may produce changes in production levels but only those affecting the genetic performance levels of the sheep being farmed are considered here.

The first important genetic decision concerns the choice of breed(s) on the farm (see Chapter 2). An associated decision is the mating system to be used (see Chapter 5). Crossbreeding, as commonly employed for grading-up or new breed formation, does not provide a continuing source of genetic gain unless new, superior breeds are constantly becoming available. As a consequence if genetic performance levels are to increase further, it is necessary for ram breeders to carry out selective breeding within each breed, whether straightbreeding or crossbreeding.

To derive the breeding objectives for sheep improvement, those animal traits which have the most effect on net income and which will respond to genetic selection need to be determined. This will require an assessment of the contributions to lifetime ewe production and the costs of this production.

LIFETIME RETURNS

In assessing the contribution of commercial animals to flock returns it is necessary to evaluate production on a lifetime basis. Survival rates and changes in production with age can have an important influence on lifetime production per breeding ewe. The extent to which ewes produce meat from lambs depends on their annual weaning performance and on the number of years of production. On the other hand for wool production the amount and value of wool produced prior to breeding age needs to be considered along with the annual wool production of ewes at later ages. The salvage value of the ewe at the end of her life in the flock also contributes to total production.

So long as the genetic response to selection does not necessitate or allow any change in stocking rate the objective can be expressed in terms of profit per ewe lifetime. The assumption is made here that this is equivalent to profit per ewe per year. This is only true if the selection system and the genetic response do not alter the age structure of the flock or the relative performance at each age. Due account must be taken of the cost of ewe replacements, whichever method is used, otherwise the expected returns from lamb sales will be overestimated.

COSTS OF PRODUCTION AND EFFICIENCY

Costs of production are an important component of net farm income or its equivalent, net income per hectare for a farm of constant grazing area. One of the major problems of including production costs in animal breeding objectives is how to allocate them to individual animals in a farming environment where the major cost is that associated with pasture production and utilisation. Consequently in defining selection objectives breeders have tended to assume that genetic responses in production per animal will result in increased net income per ha (or per unit of food intake). This assumption is implicit in objectives defined in terms of total product value per animal.

Pen-feeding and grazing experiments in Australia support the contention that animals which produce more wool are more efficient in terms of the wool produced per unit of food consumed, and that this is also true for genetic responses to wool production.[31] [34] This evidence does not however completely solve the problem because we must account for grazing conditions where costs of providing food may vary considerably according to the season of the year. These foods costs may be relatively low during periods of peak pasture production and high during periods of low pasture production in winter, or in summer drought. For peak pasture production, animals with the ability to utilise large amounts of food are likely to be more profitable, otherwise the ability to express a high output from a minimal input could be most valuable.

In terms of New Zealand lamb production, the result of several generations of selection for multiple births suggests a positive response in output per hectare.[17] Data from a stocking rate trial comparing high fertility and control Romneys are summarised in Table 6.1 and gross income as derived from per hectare production of meat and wool is compared in Table 6.2. It can be seen from Table 6.2 that there was a 7% economic advantage to the high fertility Romney, an advantage reflected in higher *net* income unless labour costs are substantially higher.

The experiment also indicated improved biological efficiency, at least as far as the objective of the Romney selection experiment was concerned, which was to increase lamb production. As can be seen from Table 6.1 the experiment achieved 36% more lambs born per ewe joined, with a correlated response of 37% more lambs weaned per ewe joined. However, information is not provided on the costs of obtaining the improved production. The emphasis which selection programmes should place on criteria of output per ewe relative to criteria which indicate the costs of obtaining improved production has not been intensively studied. More information on feed intake is required in order to maximise genetic responses in production per hectare.

SELECTION OBJECTIVES VS SELECTION CRITERIA

To avoid confusion it is necessary to make a clear distinction between the selection objectives and selection criteria. Increasing overall profit per animal per year is the selection *objective*, and this must take account of the costs and returns associated with change in the level of expression of each trait of economic importance. The selection *criteria* are those

TABLE 6.1: Ruakura Sheep Stocking Rate Trial[17]: Physical Input and Output Data for Romneys Selected for High Fertility and at Random

Trait[1]	High Fertility	Control
LB/EL	1.60	1.18
LW/LB	0.75	0.85
LW/EJ	1.03	0.75
Ewe body weight (kg)	52.6	53.4
Lamb weaning weight (kg)	17.7	20.1
Milk secretion (litres/day)	1.39	1.17
Pasture intake (kg DM/ewe/year)	665[2]	626

[1] LB = lambs born, LW = lambs weaned, EL = ewe lambing, EJ = ewe joined

[2] Indicates higher utilisation of pasture, because the two strains were at the same stocking rates.

TABLE 6.2: Production/ha for Romneys in the Ruakura Sheep Stocking Rate Trial:[17]

Production/ha	High Fertility	Control
Lamb weaning weight (kg)	390	314
value[1] ($)	127	102
Wool clip (ewes only) (kg)	63.9	73.1
value[1] ($)	78	89
Total Value ($)	205	191

[1] Unit values are recommended product-price assumptions for cost-benefit analyses, MAF:[22] FOB prices. Assumed carcass weight = 47% of live weight.

traits which are assessed in order to predict the objective.

All factors of economic importance should be included in the objective, but it is not necessary to be able to measure directly the various components of the objective.[13] Rather one should have reliable estimates of the extent to which components of the objective can be changed by selection. Then a selection system has to be applied to enable traits to be given the appropriate emphasis, so that genetic change in the combined selection objectives is as fast as possible. The selection index method is often convenient where there are a number of economically important components of the selection objective and where the selection criteria can be assessed quantitatively.

It is the aim here to highlight and take into account some of the difficulties which arise in quantitatively defining objectives for sheep improvement, and to study some of the consequences of using alternative selection objectives when the definition of the objective is not clearcut.

SPECIALISATION OF BREEDING OBJECTIVES

Various breeds of sheep are valued for different types of production functions, and have evolved both to suit a range of farming environments and for different types of products. The New Zealand sheep industry has become stratified to make use of some of these differences between breeds. The two major categories are the specialised prime-lamb-sire breeds and the dual-purpose breeds (see Chapter 2).

The dual purpose breeds can be subdivided according to wool type and wool selection objectives into:

(1) *General purpose wool breeds* such as the Romney, Coopworth and coarse-woolled Perendale.
(2) *Fine woolled breeds* such as the Merino, New Zealand Halfbred and Corriedale.
(3) *Specialty carpet wool breeds* such as the Drysdale.

The Merino is listed here as a dual-purpose breed although it is often regarded as a specialist wool producer. However in New Zealand its meat-producing function is also important.

Dual-purpose sheep are used in two main types of breeding systems:

(1) Straightbreeding or crossbreeding to generate dual-purpose breeding ewes as well as male lambs for slaughter;
(2) Crossbreeding with a specialised meat sire, to generate suitable lambs for slaughter.

Using the Romney breed as an example; in order to provide sufficient Romney ewe lambs to maintain breeding ewe numbers (see Appendix 6.1), approximately 56% of the Romney ewes must be mated to Romney rams. The remainder can be mated to produce slaughter lambs (see scheme 1, of Table 6.3) or crossbred lambs (schemes 2 or 3). It is a

TABLE 6.3: Specialisation of Breeding Objectives Among Flocks which Supply Rams to the Commercial Industry

Mating scheme	Breeding objective[2] in ram breeding flocks	Mating combination[1] in commercial flocks		Lamb breed and function
		Ram Breed	Ewe Breed	
1	A	R	R	R . . replacements and slaughter.
2	C	R	R	R . . replacements
	T	D	R	DR. . slaughter.
3	A	R	R	R . . replacements
	A	B	R	BR . . slaughter ♂, ♀ for breeding stock.

[1] R = Romney; B = Border Leicester; D = Southdown. (These are examples only: R and B could be any dual purpose breed, D could be any prime lamb sire breed).
[2] A, C and T are defined in the text.

common practice for many ewes to be mated to rams of the same dual purpose breed for several years while being run on hill country and then, as cast for age ewes on lowland country, to be mated to specialty meat sires (terminal prime lamb sire breeds). Others are used for crossbreeding throughout their life.

In flocks concentrating on crossbreeding for prime lamb production the genes of the dual-purpose breeds affect ewe wool production, fertility, fecundity and other maternal characteristics of the ewe flock, but only make up half the lamb's genes for survival, growth and dressing percentage etc. The remainder of the lamb's genes come from terminal sire breeds.

In concept it is possible to define two separate breeding objectives for dual-purpose breeds. These will be referred to as:

(A) when most ewe progeny will be used as a further generation of breeding ewes; and

(C) when ewes will be used mainly for mating with prime-lamb sire breeds with all of their lambs in turn being slaughtered.

While the sheep breeder will seldom be able to define the attention to be paid to each of the above objectives it is useful to analyse the objectives separately.

An objective defined for prime lamb sire breeds used for terminal crossing will be referenced by (T).

General-Purpose-Wool Breeds

Breeding objectives will be discussed using the Romney as an example of a general purpose wool breed. It has a role in all three mating schemes described in Table 6.3: either straightbreeding, or as a parent of crossbred breeding ewes such as Border Leicester x Romney, or as a dam of prime lambs for export. Breeding objective A is considered first and objective C later.

Romney Straightbreeding: Objective A

Objective A can be defined on a per-ewe lifetime basis as a combination of numbers of lambs weaned, lamb weaning weight, lamb dressing percentage, ewe carcass weight and greasy fleece weight. No value is attached in the objective to changing fibre diameter.[26] The lack of reliable estimates of economic values has prevented the consideration of fleece whiteness, fleece character and some other wool traits as part of the objective.

A breeding objective for Romney straightbreeding can be stated in the form of an equation:

$$H_A = a_1 NLW + a_2 WWT + a_3 D\% + a_4 ECW + a_5 GFW,$$

where,

NLW = number of lambs weaned (per ewe per year);

WWT = weaning weight (kg per lamb);

148

D% = lamb dressing percentage;

ECW = ewe carcass weight (kg per ewe);

GFW = greasy fleece weight (kg per fleece);

the 'a' values are economic values or changes in lifetime net income per unit increase in each trait each year.

Each economic value in H_A is based on the recommended product-price assumptions for cost-benefit analyses[22] to the farmer, as summarised in the bottom row of Table 6.4. Many aspects of physical production are also implied in the economic values for H_A, and the important underlying assumptions have been summarised in appendix 6.1, together with the calculations for deriving the values of a_1 to a_5. Using the values calculated in appendix 6.1, this equation becomes:

$$H_A = 23.9NLW + 0.56WWT + 0.34D\% + 0ECW + 6.42GFW$$

When the New Zealand-wide sheep recording scheme (Sheeplan)[4] was devised different economic values were used. For comparison the equivalent Sheeplan breeding objective, scaled to give greasy fleece weight the same value, is

$$H_{A_2} = 38.7NLB + 1.67WWT + 0D\% + 0HBW + 6.42GFW$$

where,

NLB = number of lambs born (Sheeplan uses NLB and not NLW);

HBW = hogget body weight.

When defining a selection objective it is usual to add together weighted values of various traits. Probably some of these traits should be combined multiplicatively, because returns from higher dressing percentages and part of the returns from heavier live lambs and higher fertility and fecundity are realised through carcass receipts.

It has not been possible to give adequate consideration to food intake as part of the objective since there are no published genetic correlations of individual grazing intake with components of animal output. An estimate however has been made (see Appendix 6.1) of the marginal costs of increased food intake for the maintenance of heavier ewes, accumulated over a ewe's lifetime and charged against the marginal returns from heavier ewe carcasses.

Generally for ewes of different productive potential the grazing pressure per ewe may not be the same. No fixed costs of acquired extra land have been included in the marginal food costs of extra production. It is also assumed that non-food variable costs are no higher in the selected flocks or their progeny. It is therefore argued that extra physical output must be realised on the existing land base, through sheep which are either biologically and/or economically more efficient. If biologically more efficient, then they will produce more from the same food. This will directly affect *gross* income and *net* income. Alternatively they could be economically more efficient in another sense, producing greater *net* income from different levels of animal production and grass production on the same land. In this case extra fertiliser costs for

149

example would be more than compensated for by a higher gross income. As indicated before, the implications of these alternative types of superiority are different, and it is not possible to determine the outcome for different fertility strains by running management trials for one strain.

Since breeding programmes are long-term, an important question is what are the effects of changes in relative prices of meat and wool on the objective. Ideally future prices and costs should be used, but well-researched assessments of these are not available. In order to show the size of price changes in the past, five-yearly FOB averages are summarised in Table 6.4 for the years 1955-1975. The wool/lamb price ratios have declined from 2.35 in the first period to 1.63 in the most recent period, a relative change of 31%.

TABLE 6.4: Average Auction/Works Price for Wool and Lamb, and Price Ratios

Category	Wool price[25] c/kg	Lamb price [25] c/kg	Wool/lamb price ratio
1955-60	81.9	34.8	2.35
1960-65	80.7	33.2	2.43
1965-70	62.0	41.1	1.51
1970-75	99.0	60.9	1.63
MAF[22]FOB price	122	69	1.77
MAF[22] Farm price	107	47	2.28

To show the effect of different relative prices in the objective, a straightbreeding objective appropriate to 1955-60 farm gate prices (H_{AO}) has been calculated to compare with the straightbreeding objective using 1975 prices (H_A). The percentage and total monetary gain from using selection indexes chosen to maximize gain in straightbreeding objectives H_{AO} and H_A are shown in Table 6.5. It can be seen from Table 6.5 that the traits of most importance are NLW, GFW and WWT, in that order. The percentage of overall monetary gain derived from changes in NLW was calculated to be 64% for H_A and 47% for H_{AO}, and corresponding percentage values derived from changes in GFW were 26% and 45%. The relative emphasis on NLW and GFW in the objective can change quite considerably when wool/lamb price ratios change and hence relative economic values used in constructing an index are altered.

A better way of showing the effects of different relative economic values is in terms of expected genetic changes as in Table 6.6. This shows the expected genetic change in annual production for the traits of the objective after five generations of selection. The phenotypic and genetic parameters used in calculating these responses are given in Appendix 6.2. It is assumed that the selected sheep are one standard deviation superior in index score. This could be achieved if the top 38% of ewes

150

TABLE 6.5: Percentage and Total Monetary Gain from Using Optimal Indexes Applicable to Different Relative Economic Values in the Objective

Trait	Index to predict objective:		
	H_{AO}	H_A	H_C
NLW	46.9	64.0	67.2
WWT	9.2	11.4	4.9
D%	−1.5	−1.2	−0.7
ECW	0	0	0
GFW	45.4	25.9	28.6
Total gain ($/lifetime)	0.98	1.51	1.44

TABLE 6.6: Expected Genetic Change in Annual Production in Romneys After Five Generations of Selection, Given Objectives H_{AO}, H_A, and H_C

	H_{AO}	H_A	H_C
NLW	0.17	0.20	0.20
WWT (kg live weight)	1.41	1.53	1.25
D% (of lambs)	−0.38	−0.26	−0.29
ECW (kg carcass weight)	1.60	1.59	1.41
GFW (kg)	0.42	0.30	0.32

were used as replacements and if the rams used were of equivalent genetic merit to the ewes. Alternatively the selection intensity could be achieved by using rams from the top 6%, with no selection of ewes.

As can be seen from Table 6.6, a higher index weighting for fleece weight has favoured a greater fleece weight response for H_{AO} (an approximate 30% increase) but the expected responses in NLW, WWT and ECW are not greatly changed, considering the size of the change in relative prices. The larger expected change in dressing percentage is not economically important since genetic change in this trait only accounted for a small proportion of the monetary response. Also this is likely to be compensated for by a change in the wool pull of the lamb at slaughter.

Romney Crossbreeding: Objective C
If the breeding objective is to breed dams of prime lambs the objective is given by:

$$H_C = a_1NLW + \tfrac{1}{2}a_2WWT + \tfrac{1}{2}a_3D\% + a_4ECW + a_5GFW$$

The coefficients of one half for WWT and D% represent the appropriate genetic contribution of the Romney to the crossbred lamb production traits[33]. Some would question the value of ½ as the coefficient for WWT, in view of the effects of maternal factors such as the dam's milk production. The coefficient of NLW (a_1) is taken here in units of "dollars per ewe lifetime per unit increase in NLW". NLW is a composite (multiplicative) trait, and is made up of variation in ewe survival, ability to get in lamb, litter size and lamb survival. Of the four parts of NLW, the first three are usually largely determined by the ewe, while the influence of the ram's genes is more readily seen in the last component, lamb survival.

The percentage response expected in each trait and the total monetary gain for the crossbreeding objective (H_C) using an optimal selection index are shown in Table 6.5 to enable comparison with the straightbreeding (H_A). For H_A and H_C the percentage response expected in each trait is similar (Table 6.5). The largest difference is for WWT, a reduction (11.4% to 4.9%), because much of the improvement in lamb weights in the crossing programme will come from the specialised meat sire.

Table 6.6. shows the expected genetic changes in the various traits resulting from selection for the different objectives. Use of the index for the objective H_C in comparison with that for H_A is expected to result in a 7% greater response in fleece weight, an 18% smaller response in weaning weight and an 11% smaller response in ewe carcass weight. There was no important change in the expected response in NLW and D%.

The general conclusion is that differences between genetic changes following selection for H_A and H_C (as shown in Table 6.6) are probably not large enough to justify the use of two objectives for coarse-woolled dual-purpose breeds. Although it is technically possible to use two objectives (and therefore two sets of indexes) within the one recording scheme, confusion could arise amongst breeders. Therefore, inclusion of additional indexes in recording/selection systems such as Sheeplan seems unnecessary.

Summary of Objectives for General-Purpose-Wool Breeds

For breeds like the Romney, important points are:
(1) The major objectives are *numbers of lambs weaned, fleece weight* and *weaning weight* (in that order);
(2) Index selection can help the breeder select for all three at once;
(3) The relative emphasis on the numbers of lambs weaned and greasy fleece weight in the objective can change quite considerably when wool/lamb price ratios change and hence the relative economic values used in constructing an index are altered;
(4) It would probably not be worthwhile attempting to include separate

straightbreeding and crossbreeding objectives for each dual-purpose breed. The distinctions between the two types of objective are small in terms of genetic changes in practice.

Fine-Woolled Breeds

Merinos, Corriedales, New Zealand Halfbreds and the finer woolled Perendales have breeding objectives essentially similar to other dual-purpose breeds, as income is derived from both meat and wool. However, aspects of wool quality are of greater importance than for general-purpose-wool breeds.

Until recently quality number was the main commercial criterion of wool fineness. This is being replaced by visual estimates and measurements of fibre diameter. The textile industry now recognises that mean fibre diameter is of overwhelming importance in the manufacture of finer apparel yarns.

In New Zealand, wool prices have been shown to increase with increasing quality number for wool finer than 48/50s or 34 microns.[15] [26] In Australia, where there is a predominance of fine apparel wool from Merinos, quality number was shown to be the main determinant of wool prices.[21] [23] This was used to justify consideration of quality number as a selection objective. The recent change in evaluation methods has led to fibre diameter replacing quality number in selection objectives and Australian breeders are being advised to select for greater fleece weight with an upper limit to fibre diameter.[36] [37] [38]

The change in evaluation within New Zealand also indicates that the objective should be defined in terms of fibre diameter rather than quality number. When differences between sheep within flocks are analysed there is not a close relationship between fibre diameter and quality number.[41]

Continued direct selection for fleece weight can lead to economically undesirable correlated changes in wool fineness. The importance of such changes depends on relative economic values which may not remain constant over time. Unfortunately it is possible for the industry to change the relative economic value of a quality trait quickly, as witnessed by the price advantage of finer wools which fell dramatically in 1976/77.

The breeding objective for fine-woolled breeds may be written as:
$$H_F = a_1NLW + a_2WWT + a_3D\% + a_4ECW + a_5GFW + a_7SL + a_8FD,$$
where a_7 and a_8 are economic values or changes in lifetime net income per unit increase in staple length (SL in cm) and fibre diameter (FD in microns), respectively. The value for a_8 is negative.

A similar objective has already been analysed for Perendales with a different set of economic values, expressed in terms of net income change for each unit change per ewe per year[10] (See Table 6.7). There

153

were five traits in the selection index, NLW, GFW, HBW, SL and FD.

The expected genetic changes following five generations of selection are given in Table 6.7, along with the contribution which each trait of the objective makes to the total monetary gain expected. The most important traits are NLW, GFW, and SL, in that order. The use of a reduced index which included only NLW, GFW, and HBW was likely to be 14% less effective. The cost of measuring staple length and particularly fibre diameter needs to be compared against the extra return expected from a more detailed index. This was not done, but it was suggested that the extra improvement was sufficient to warrant inclusion of fibre diameter and staple length in the selection index, particularly for selection of Perendale rams in important ram breeding flocks.

TABLE 6.7: Selection Objective and Expected Genetic Gain, from an Index Calculated for Perendales[10]

	Economic value ($)[1]	Genetic gain[2]	% of total $ gain
NLW (lambs weaned/yr)	7.00	0.19	53.0
GFW (kg)	1.50	0.40	24.3
HBW (kg)	0.00	0.50	0
Staple length (cm)	0.25	1.76	18.0
Fibre diameter (μm)	−0.20	−0.59	4.8

[1] Per ewe year, per standard deviation.

[2] For five generations with one standard deviation of selection intensity on the index. This is comparable with Tables 6.5 and 6.6 for coarse-woolled sheep.

One result of using the reduced index (NLW, GFW and HBW) was that following five generations of selection the fibre diameter was expected to be 1.0 micron coarser compared with 0.6 microns finer with the full index.

The consequences of using HBW instead of ECW are probably small because the economic values have been estimated as zero, and changes in components of the objectives due to selection were not sensitive to differences in the economic value of HBW.

How far the results for Perendales can be extrapolated to Corriedales and other breeds, particularly Merinos, is not known. Genetic parameters may be different. Average performance levels of Corriedales and Merinos are different from Perendales, and this affects the coefficients used as economic values in the selection objective. In some Merino flocks there are only enough females born to replace older ewes. Some of the implications of different economic values for Merino selection can be obtained from Australian studies.[35, 42] NLW is still the major objective.

Summary of Objectives for Fine-Woolled Breeds

(1) The breeding objectives for fine woolled breeds are similar to the general-purpose-wool breeds, but with the addition of staple length and fibre diameter.
(2) *Numbers of lambs weaned* contributes over 50% of the monetary value of genetic changes expected from selection for a Perendale objective.
(3) *Fleece weight* is the most important wool characteristic.
(4) *Staple length* is more important in the appropriate selection index for Perendales than *fibre diameter.*
(5) It would be necessary to obtain estimates of the parameters applicable to New Zealand Halfbreds, Corriedales, and Merinos if objectives along with their selection indexes are to be accurately assessed for these breeds under New Zealand conditions.

Specialty Carpet-Woolled Breeds

The wool produced by the Drysdale, Tukidale and Carpetmaster breeds is described as specialty carpet-wool, because it is heavily medullated (hairy), bulky and springy.[6] These breeds must be recognised as dual-purpose breeds producing meat and specialty carpet wool. The breeding objectives are similar to those of Romneys except that aspects relating to specialty carpet wool production must be added. Future market requirements will influence the relative importance of meat and specialty wool as products. As with Romneys, the priority is high fertility as this is the key factor influencing returns, and predetermining the amount of culling and therefore genetic progress in other characteristics. To increase meat production, the objective also includes weaning weight, lamb dressing percentage and ewe carcass weight. As in the other wool breeds, fleece weight is an important productive character, for it largely determines the value of wool per sheep within the breed. It is also expected that sheep which produce the heaviest fleece weights will be the most efficient biologically.

When considering what emphasis should be put on improving the wool quality of fleeces from specialty carpet-woolled breeds, a clear understanding is required of the end use of this type of wool.

Carpets are extremely variable in nature. The quality of the carpet is determined by the material used in the pile, its backing and by the method of manufacture. Wool which is incorporated into the pile is made from a blend of fibres (whether the blend is all-wool, or a combination of wool and non-wool fibres). Its formation is usually a compromise between cost and performance. The careful selection of material for blends is probably the most important single factor in achieving good processing and end product quality.

Wool used in carpets has been classified into three main categories:[30]

(1) *Traditional types*, heavily medullated.

(2) *Crossbred types*, lightly medullated.

(3) *Filler wools*, variably medullated.

The specialty carpet wools of Drysdales, Tukidales and Carpetmasters are included in category (1). These wools are used almost exclusively in carpets and normally constitute 15-20% of the carpet blend.[20] Performance trials[31] have shown that carpets containing 40% of specialty carpet wool have not performed as well as those containing 20%, although satisfactory carpets of 100% Drysdale have been tested.[24] When specialty carpet wools were used in only small percentages in a blend, (less than 10%) no specific conclusions could be drawn about their performance.

Carpet manufacturers cannot provide the specialty carpet-wool breeder with exact specifications of medullation, bulk, resilience, staple length, lack of lustre and harshness etc. Hence the breeder still finds it difficult to determine what components of wool quality should be included in the objective for improvement. In the short term only those for which a premium is paid warrant attention as objectives, but the breeder may be mindful of future trends in the wool industry.

To be included in a breeding objective, a character must be of economic importance. It must also be able to respond to selection. No genetic parameters have been estimated for the Drysdale, Tukidale or Carpetmaster breeds. It is not known whether the fibre properties will respond to selection, and if so, what will be the pattern of this response in relation to those of other productive characters. Consequently, it is not yet possible to formulate a selection index and hence make a clear definition of the most effective use of selection for improving wool quality in a specialty carpet wool breed.

Selection guidelines for specialty carpet wool breeds have however been published.[26] These guidelines were:

(1) *major selection characters:* fertility, fleece weight, greater medullation

(2) *other criteria:* bulk, whiteness, freedom from lustre, growth rate.

While the broad principles for improvement in sheep production (i.e. fertility, fleece weight and growth rate) are known, specific details of how improvement in the fibre can be accomplished are not clear. Before any productive character can be considered as a criterion, it must be capable of being assessed accurately. Medullation may be assessed by eye, but rather inaccurately, and it can be measured,[32] as can bulk which is highly correlated with resilience[9] and wool colour (whiteness).[14] At present most carpet-wool breeders select sheep for greater medullation as assessed by eye. However insufficient information is available to estimate the relative economic importance of these characters at present, so the importance of growth rate relative to the specialty wool traits is not known.

Summary of Objectives for Specialty-Carpet-Woolled Breeds

(1) Objectives of major importance in these breeds are *fertility, fleece weight, lamb carcass weight,* and *greater medullation.*

(2) The economic values of most wool quality traits are still not known.

(3) Genetic parameters have not been calculated for traits of carpet-wool breeds.

(4) *Medullation, bulk,* and *whiteness* could be incorporated in some recording schemes in the future if economic and genetic parameters justify it. The costs of measurement and recording are expected to be high, but the traits may have their place for the selection of rams in important ram breeding flocks.

Prime Lamb Sire Breeds

The breeding objectives for prime lamb sire breeds, such as the Southdown, are in general terms a combination of lamb survival of crossbred progeny, lamb sale weights and carcass value.[1][8]

Breed of sire can have an important effect on progeny weights.[1] A major consideration in the choice of breed for use as a terminal sire is the fact that the New Zealand lamb price schedule classifies lambs by weight in three distinct categories. These categories have had large price differentials in the past which discouraged the production of heavy carcasses.[28] However there are indications that in the future the production of lean heavy-weight lamb carcasses will be encouraged. There appear to be improving opportunities to achieve financial rewards by increasing carcass weights provided fat cover can be kept within strict limits.

While *wool-pull* payments can be an important aspect of returns from slaughtered lambs there is very little information on factors affecting them. Lamb fleece weights will be ignored here, so economic gains from genetic selection within breeds will be slightly underestimated because the genetic and phenotypic correlations of hogget fleece weight with weaning weight and hogget body weight are positive within breeds. This may be balanced partly by a strong negative correlation between lamb fleece weight and dressing percentage.

The objective (H_T) has already been defined tentatively in relation to crossing prime lamb sires to Romney ewes and it may be written as:

$$H_T = \tfrac{1}{2}\, a_2\, WWT + \tfrac{1}{2}\, a_3\, D\%$$

More realistically the objective should be defined multiplicatively as:

$$H_T' = \tfrac{1}{2}\, a_6\, (LS \times ALW \times \frac{D\%}{100})$$

where,

LS= lamb survival (birth to slaughter);

ALW = lamb live weight at slaughter (genetic parameters are assumed to be the same as for autumn live weights of breeding animals);

a_6 = lamb price/kg carcass weight (within the 13.0-16.0 kg range). Ideally this should be corrected for the costs of extra food intake of heavier animals at the same age (a_6 here = $0.47/ kg carcass).

TABLE 6.8: Expected Genetic Change in Annual Production in Crossbred Lambs from Southdown Sires after Five Generations of Selection, Given Objective H_T'

Parameter	Genetic Change[1]
LS (%)	5.50
ALW (kg)	3.81
D (%)	0
Gain (in $/ewe mated)	0.042 (Index = corrected WWT)
Gain (in $/ewe mated)	0.105 (Index = corrected WWT and ALW)

[1] Assumes the same selection intensity as in Tables 6.6 and 6.7. (See text also).

The consequences of using breeding objectives H_T' are described in Table 6.8, with estimates of the genetic changes expected after five generations of selection. As for the Romney and Perendale objectives, it is assumed that the average of the selected sheep is one standard deviation of index score better than that of the original group before selection. The index in this case is a linear combination of corrected WWT and ALW.

Two other indexes were also compared with the linear combination: corrected WWT and corrected ALW. Because of the coefficient of ½ in H_T' the economic data refer to the extra net income from crossbred lambs sold, i.e. extra income due to selection in Southdowns. The extra net income in the same lambs due to selection in the Romney parent was given for H_C in Table 6.5 as $1.44/ewe lifetime (=$0.38/ewe mated per year). Thus selection in Romneys is 4 to 9 times more rewarding than selection in Southdowns, in terms of monetary gains expected; an important contrast between the selection responses in dual-purpose and prime-lamb sire breeds. The reason for the difference is that ewe fertility contributes greatly to the dual-purpose objective, but is not part of the prime-lamb sire breed objective. In practice however the Southdown selection intensity could be higher than the one standard deviation assumed for Romneys, and the generation interval could be shorter.

These two factors could lessen the difference between the economic selection responses of the two breeds.

Lamb survival (LS) and autumn live weight (ALW) are expected to contribute 33% and 67% respectively to genetic gain in net income in the prime lamb breed objective (H_T'). Such calculations depend on the genetic parameters assumed for traits in the objective and the indexes. The phenotypic and genetic parameters used to construct selection indexes for the objective H_T' are given in Appendix 6.3. There is little information on these parameters, particularly for dressing percentage, and correlations of zero with D% have been assumed here.

The estimates used here show that within-breed genetic change in ALW is about twice as important as LS. More progress in LS might be possible if more information was available about the genetic control of lamb deaths, enabling direct selection for a more heritable component of LS.

Summary of Objectives for Prime-Lamb-Sire Breeds

1. The trait of major importance for selection is *live weight at sale*, at least when carcasses are still below 16 kg.
2. *Lamb survival* is about half as important as *autumn live weight*.
3. Not enough is known about genetic parameters for *dressing percentage* and other carcass traits.
4. The rewards per animal from selection for commercially important traits in prime lamb breeds are substantially less than in dual-purpose breeds.

Additional Components of the Breeding Objectives

It is pertinent to ask what information was used to decide on traits in the breeding objectives. This is an area where neither the principles nor the practice are well documented. The breeding objective should contain those *animal traits* (in contrast to *managerial factors*) which have the most important effect on net income. It is often difficult to decide on the dividing line between animal traits and managerial factors, especially for disease resistance, all-or-none traits such as barrenness, particulate traits such as litter size (1, 2 or 3 lambs), and aspects of structural soundness. Disease resistance and structural soundness are generally ignored by scientists in attempting to define breeding objectives quantitatively*. This is because of ignorance about correlations between these traits and other aspects of production, like fleece weight or growth rate. The lack of information is largely due to difficulties in measuring the traits consistently. The consequences of ignoring them are not known.

*Note:- This statement does not imply that scientists ignore the presence of structural unsoundness when selecting breeding stock. On the contrary strict attention to culling animals with obvious structural defects of the teeth, pasterns, feet and testicles is practised.[16]

Sheep Breeding and Reproduction

The question is not whether managerial factors affect such traits, but does the level of management affect the *ranking* of animals for net income per ewe? In other words are *genotype × environment interactions* important? The answer is not known, but is likely to depend on the size of managerial and environmental differences (e.g. climate, terrain, stocking rate, lambing date, drafting date, degree of shepherding).

Structural Soundness

The importance of soundness in physical characteristics associated with production (feet, jaws, teeth) and reproductive organs has already been mentioned. There is considerable variance of opinion in the industry as to how far from the breed average any of these aspects can stray before sheep become structurally unsound to the detriment of net income per ewe. Also little has been published on how structural soundness changes with age, and on how it affects later production, although much of the culling of females in the industry is done on age, soundness and condition. The economic effects of these structural deformities have never been adequately analysed.

Disease Resistance

For facial eczema (FE), evidence is accumulating that there are differences among progeny groups for lamb survival which are partly attributable to genetic differences in resistance to FE.[2] The incidence of FE is confined to certain geographical areas of the North Island. More needs to be known about its heritability and genetic correlations before it could be included in the breeding objective defined in H_A. Methods of testing young rams for susceptibility are being investigated. In terms of the possibility of measuring FE incidence in progeny groups for dual-purpose breeds, the time of the year when stock are at risk occurs after weaning. Care must therefore be taken to avoid biasses in progeny tests due to culling (precisely the same problems as in all sire summaries).

More information is needed on possible genetic resistance of sheep to FE and to many other diseases of economic importance. For any diseases where some resistance is conferred genetically, it is not known for the long term whether the best control method will be by genetic or managerial procedures. That is to say, should we select sheep for genetic resistance to FE, or should we spray pastures or use different grazing methods?

Other Attributes of Breeding Stock

Some aspects of production which have not been mentioned yet are *easy-care, open-faces, hardiness,* and *free-moving stock.* Some argue that these attributes are incorporated into number-of-lambs-weaned or live-weight traits and hence are automatically selected in a performance-backed breeding programme. In commercial flocks they do not

160

contribute to extra gross income, except through extra weight of lamb or fleece sold per flock-year, and these two traits have been defined specifically as components of the breeding objective. Potentially they could cause decreased costs (and hence increased flock net income) through less shepherding costs at lambing, mustering etc.

Wool blindness can be a problem and many commercial farmers will not buy woolly-faced rams. This has resulted in open faces becoming an objective for many ram breeders.[16] Face cover score has been used as one of the predictors of higher fertility;[5] it can therefore constitute a component of the selection index.

Selection or culling on appearance may include some or all of the following criteria: *body size, shape, fleece characteristics, constitution* and *body condition.* It should be noted however that they are only criteria, rather than objectives for improvement of progeny. If there is also a correlation between sale price of rams and either actual weight or condition, then this would indicate that commerical producers and other breeders attach importance to them. Only the part of variation in these traits which is genetically correlated with extra weight of lambs or fleece sold per ewe-year is important for improving progeny. Visual assessment of fleece production also qualifies in this way as a selection criterion, although measured fleece weight is 3-4 times more accurate.[29] In addition selection on body size and body condition could affect current flock production levels favourably (i.e. through a phenotypic response).

Other attributes sometimes used as criteria for culling in some breeds are *black spots, horns/scurs* and *hairiness.* Their economic importance has not been defined clearly. Insufficient is known about the incidence of black spots.[7] Generally the aim is to cull animals to keep breeding stock free from these undesirable characteristics.

Fleece character and *style grade,* two subjectively assessed characteristics, have in the past been used as criteria for culling. Fleece character has been shown to have no influence on price,[8 23] to have a low heritability in hoggets and to be poorly related to the important wool production traits.[12 27] Because of the small size of correlations between the character of hogget fleeces and ewe traits in later life and the fact that *lifetime ewe production* has been found to have a negative association with hogget fleece character in Perendales it has been suggested that the common practice of culling hoggets with low character grades cannot be justified.[11] Consequently its inclusion in an objective for sheep improvement seems unwarranted.

Style grade is a composite of economically important traits which can be objectively measured, such as *whiteness, staple strength* and *vegetable matter content,* as well as subjectively assessed traits such as the *degree of skirting* and *evenness* (or conversely the irregularity) of length and quality. These traits are strongly influenced by or may be solely dependent on management factors. While unquestionably style grade influences wool prices, current research aims to determine the economic importance and inheritance of those objectively measurable traits which make up style grade. Also the relationships between style and other productive traits are unknown. For these reasons style grade is ignored in the objective.

Other traits currently of high interest in sheep breeding research include the number of hogget oestruses,[3] testis diameter,[19] and early sexual maturity.[16] All may be used in the future as selection criteria to help predict reproductive rate.

Relationships Between Ram Breeding and Commercial Flock Objectives

The objectives already defined are those appropriate to all the sheep within each grouping. However traditionally the sheep industry is stratified with most sheep breeders relying on rams purchased from a relatively small proportion of the flocks (See Chapters 2 and 7).

While simple, low-cost selection procedures can produce sufficient response to more than pay for the costs of gaining some improvement in fertility[34] and fleece weight,[35] the use of more sophisticated procedures financed by returns from ram sales will allow faster rates of progress. Thus the segregation of the population into "ram-breeding" and "commercial" flocks is justified. Within commercial flocks the rate of genetic improvement will be largely determined by the rate of genetic improvement in the flocks supplying rams.

Unfortunately a selection objective designed to maximise profit in a ram breeding flock may not be the best for improving the profitability of the commerical farms which use the rams.

A problem with the traditional system of purchasing is that ram buyers have relied heavily on visual assessment of rams. This in turn has made visual characteristics of rams of considerable economic importance to ram breeders. In consequence they have tended to emphasize conformation and visually-assessed wool characteristics to the detriment of productive traits of higher economic importance to commercial farmers. Clearly ram buyers must disregard purely cosmetic factors in ram assessment if they wish ram breeders to avoid considering factors not required in commercial sheep.

Ideally ram buyers should seek to buy rams from flocks where the average genetic level (defined in terms of the commercial objective) is high. Also they should be prepared to pay more for rams of higher breeding value,[40] particularly those born within flocks of high average genotype. This would encourage ram breeders to use their records efficiently in order to raise the average genotype of their flock. However there are problems in comparing the genetic merit of different flocks since most differences in average levels of production between flocks appear to be due to differences in the environment in which various flocks perform. Perhaps practical systems of demonstrating genetic differences between flocks should be developed although systems appropriate to large mammals tend to be complex and costly.

CONCLUSION

Objectives for sheep improvement have been defined for four major classes of breeds in this chapter. These objectives tend to reflect the fact that the broad principles of improving fertility, fleece weight and growth rate are known.

Gaps are evident in the knowledge of aspects of food intake (and consequently efficiency of production), structural soundness and disease resistance.

Future market requirements can be expected to influence the relative importance of meat and wool, and as well, meat and wool quality traits. Reliance is placed on those in marketing and processing research to demonstrate relative economic values of objectively measurable traits.

Knowledge of the inheritance of many traits is limited, particularly for traits of recent interest and for the newer and lesser breeds of sheep in New Zealand. It is the responsibility of geneticists to both assess the need for the provision of this information and seek it if required.

Research workers are continually adding to the available knowledge in many fields affecting sheep improvement. This, together with changing economic circumstances, leads to a continuing need to review the breeding objectives and the methods of selecting sheep.

It is important that commercial producers are judicious in their choice of flock sires and that the methods of choosing sires encourage ram breeders to performance record and carry out selection procedures for commercially important productive traits.

It is for the scientists and farm advisors to demonstrate effectively to the whole industry the economic merits and relevance of objective selection procedures. It is for the sheep industry to make effective use of relevant information and techniques (such as Sheeplan) to obtain widespread adoption of genetic improvement.

APPENDIX 6.1: Underlying Productivity Assumptions, and Calculations of Economic Values for H_A

1. Proportion of two-tooths in the flock at lambing = 0.265.
   Therefore number of matings per lifetime (NML) = 1/0.265 = 3.77 assuming no significant contribution from mating ewe lambs.

2. Years of fleece production per lifetime (up to final lambing) = 3.77 + 1.
   Assuming ewes are culled 6 months after lambing, add 0.5 years.
   Crediting the fleece production of culled ewe hoggets to ewes' life-time fleece production add 0.77 years, assuming all ewe hoggets are wintered (i.e. (½ × 0.94 − 0.265)/0.265). (See below for 0.94).
   Therefore grand total annual expressions of fleece production per ewe lifetime (NFL) = 6.04.

3. National average lamb weaning percentage = 94%.[25]
   Total lambs weaned per ewe lifetime = 0.94 × 3.77 = 3.54.
   Number of female replacements required/ewe at constant population size = 1.
   Therefore number of lambs on which returns from higher weaning weights or dressing percentages are based (NSL) = 2.54.
   Percentage of matings required for flock replacement = 100/(½ of 3.54) = 56%.

4. Average lamb carcass weight (LCW) = 13.5 kg
   Dressing percentage: lambs (D%) 47%
                          ewes (D_e%) 45%
   Ewe carcass value/kg = 60% of lamb carcass value/kg.

5. Ewe marginal food cost (per kg live weight per year) = $0.0272, (derived from that part of the variable cost which is food-related ($2.00/stock unit) and assuming that 1 kg of extra body weight on a 55 kg ewe is equivalent to a +1.36c food maintenance requirement:

$$\frac{1.36}{100} \times \$2.00).$$

This is equivalent to stating that, for a ewe of weight W, food for maintenance is proportional to $W^{0.75}$
Therefore ewe marginal lifetime food cost = 4.77 × $0.0272/kg live weight = $0.13 per kg live weight (i.e. from its own weaning to slaughter).
a_1 to a_5 are determined as follows for Romneys:

a_1 = NML × LCW × lamb carcass price per kg (LCP)
  = 3.77 × 13.5 × 0.47
  = $23.92 per extra lamb weaned per year

a_2 = NSL × (D%/100) × LCP
  = 2.54 × 0.47 × 0.47 = $0.56 per kg lamb live weight at weaning

a_3 = NSL × (slaughter live weight) × LCP ÷ 100

    = 2.54 × (13.5 × $\frac{100}{47}$) × 0.47 ÷ 100 = $0.34 per 1% D%

a_4 = (0.6 LCP) − (food cost per kg live weight ÷ D_e)

    = (0.6 × 0.47) − (0.13÷ 0.45) = $0.00 per kg ECW

a_5 = NFL × wool price per kg

    = 6 × 1.07 = $6.42 per kg greasy fleece weight per year

APPENDIX 6.2: Phenotypic and Genetic Parameters[1] for H_A and H_C (tables 6.5, and 6.6)

	Phenotypic st. dev'n	NLW[3] (dam trait)	WWT	D%	ECW	GFW
NLW (lambs/yr)	0.65	0.10[4]	0.12	0.00	0.20	−0.05
WWT (kg)	3.00	0.12	0.20	0.00	0.70	0.20
D%	4.00	0.00	0.40	0.35	0.00	−0.10
ECW[2] (kg)	5.50	0.15	0.50	0.40	0.35	0.30
GFW (kg)	0.45	0.00	0.30	0.00	0.40	0.30

[1] For explanation of symbols, see text.
[2] These were derived from ewe live weight data, assuming 0.45 for the standard deviation.
[3] Repeatability = 0.15.
[4] Heritability underlined on the "diagonal"; genetic correlations above and phenotypic correlations below the diagonal.

APPENDIX 6.3: Phenotypic and Genetic Parameters[1] for H_T' (Table 6.8)

Traits in H_T'	Phenotypic st. devn	Heritability	Genetic correlations WWT	ALW
LS (= $\frac{\text{lambs weaned}}{\text{lambs born}}$)	0.357	0.10	0.12	0.20
D%	4.00	0.35	0.00	0.00
ALW (kg)	3.40	0.22	0.70	
WWT[2] (kg)	3.10	0.10		0.62[3]

[1] For explanation of traits, see text.
[2] Not part of the objective: phenotypic and genetic parameters for WWT are required in order to calculate covariances for index traits.
[3] Phenotypic correlation.

REFERENCES

[1] Carter, A. H.; Kirton, A. H.; Sinclair, D. P. 1974: Sires for export lamb production. 1. Lamb survival, growth rate, and wool production. *Proc. Ruakura Frs' Conf.:* 20-28.

[2] Campbell, A. G.; Mortimer, P. H.; Smith, B. L.; Clarke, J. N.; Ronaldson, J. W. 1975; Breeding for facial eczema resistance? *Proc. Ruakura Fmrs' Conf:* 62-64.

[3] Ch' ang, T. S.; Rae, A. L. 1972: The genetic basis of growth, reproduction, and maternal environment in Romney ewes. II. Genetic covariation between hogget characters, fertility, and maternal environment of the ewe. *Aust. J. Agric. Res. 23:* 149-165.

[4] Clarke, J. N.; Rae, A. L. 1977: Technical aspects of the national sheep recording scheme (Sheepplan). *Proc. N.Z. Soc. Anim. Prod. 37:* 183-197.

[5] Cockrem, F. R. M.; Rae, A. L. 1966: Studies of face cover in New Zealand Romney Marsh sheep. I. The relationships between face cover, wool blindness, and productive characters. *Aust. J. Agric. Res. 17:* 967-974.

[6] Dalton, D. C.; Bigham, M. L.; Wiggins, L. K. 1973: Specialist carpet-wool sheep. *Proc. Ruakura Fmrs' Conf:* 21-33.

[7] Dalton, D.C.; Bigham, M. L.; Elliott, K. H. 1977: Black spots in sheep. *N.Z. Jl. Agric. 134* (5): 26-27.

[8] Dunlop, A. A.; Young, S. S. Y. 1960: Selection of Merino sheep: An analysis of the relative economic weights applicable to some wool traits. *Emp. J. Exp. Agric. 28:* 202-210.

[9] Dunlop, J. I.; Carnaby, G. A.; Ross, D. A. 1974: Bulk. Part I. The bulk of loose wool. *Communication No. 28, Wool Research Organisation of N.Z.*

[10] Elliott, K. H.; Johnson, D. L. 1976: Selection indices for Perendale sheep. *Proc. N.Z. Soc. Anim. Prod. 36:* 23-29.

[11] Elliott, K. H.; Rae A. L.; Wickham, G. A. 1979: Analysis of records of a Perendale flock. I. The Phenotypic relationships of hogget characteristics to lifetime performance. *N.Z. Jl Agric. Res. 22:* 259-265.

[12] Elliott, K. H.; Rae, A. L.; Wickham, G. A. 1979: Analysis of records of a Perendale flock. II. Genetic and phenotypic parameters for immature body weights and yearling fleece characteristics. *N.Z. Jl Agric. Res. 22:* 267-272.

[13] Gjedrem, T. 1972: A study of the definition of the aggregate genotype in a selection index. *Acta Agric. Scand. 22:* 11-16.

[14] Hammersley, M. J.; Thompson, B. 1974: Wool colour measurement. *Communication No. 27, Wool Research Organisation of N.Z.*

[15] Henderson, A. E. 1969: Wool quality and relative prices. *Proc. Ruakura Fmrs' Conf:* 26-35.

[16] Hight G. K.; Gibson, A. E.; Wilson, D.A.; Guy, P. L. 1975: The Waihora sheep improvement programme. *Sheepfmg A.:* 67-89.

[17] Joyce, J. P.; Clarke, J. N.; MacLean, K. S.; Lynch, R. J.; Cox, E. H. 1976: The effect of level of nutrition on the productivity of sheep of different genetic origin. *Proc. N.Z. Soc. Anim. Prod. 36:* 170-178.

[18] Kirton, A. H.; Carter, A. H.; Clarke, J. N.; Sinclair, D. P.; Jury, K. E. 1974: Sires for export lamb production. 2. Lamb carcass results. *Proc. Ruakura Fmrs' Conf:* 29-41.

[19] Land, R. B.; Sales, D. I. 1977: Mating behaviour and testis growth of Finnish Landrace, Tasmanian Merino and crossbred rams. *Anim. Prod. 24:* 83-90.

[20] Larsen, W. A. 1974: Breeding of sheep for carpet wool production — a manufacturer's viewpoint. *Wool 5 (5):* 21-25.

21 McKinnon, J. M.; Constantine, G.; Whiteley, K. J. 1974: Price determining characteristics of greasy wool. Part I: General study. In *Objective Measurement of Wool in Australia*. (Eds M. W. Andrews and J. G. Downes) Australian Wool Corporation, Melbourne.

22 M.A.F. 1975: Cost benefit analysis product price assumptions, August 1975. Technical Paper No. 4/75, Resource Economics Section, Economics Division, Ministry of Agriculture and Fisheries, Wellington.

23 Mullaney, P.E.; Sanderson, I. D. 1970: Relative economic importance of some wool quality traits for Merino and crossbred wool types. *Aust. J. Exp. Agric. Anim. Husb. 10*: 544-548

24 Nash, C. E. 1964: The assessment of N-type fleeces. II. The physical properties and practical testing of the finished carpets. *J. Text. Inst. 55*: T309-T323.

25 N.Z.M.W.B.E.S. 1976: *Annual review of the Sheep and Beef Industry* (1975/76): Publ. No. 1753, New Zealand Meat and Wool Boards Economic Service, Wellington. (Also Publ. No. 1400, for 1965/66).

26 N.Z.S.A.P. 1974: *Guidelines for Wool Production in New Zealand*. Occasional Publ. No. 3: New Zealand Society of Animal Production.

27 Rae, A. L. 1958: Genetic variation and covariation in productive characters of New Zealand Romney Marsh sheep. *N.Z. Jl Agric. Res. 1*: 104-123.

28 Rattray, P. V.; Drew, K. R.; Moss, R. A.; Beetham, M. R. 1976. Production of heavyweight lambs. *Proc. Ruakura Fmrs' Conf.*: 27-33.

29 Riches, J. H.; Turner, H. N. 1955: A comparison of methods of classing wool flocks. *Aust. J. Agric. Res. 6*: 99-108.

30 Ross, D. A. 1970: The Drysdale. *Report No. 3, Wool Research Organisation of N.Z.*

31 Ross, D. A. 1978: The characteristics of carpet wool in relation to processing and performance. *Proc. N.Z. Soc. Anim. Prod. 38*: 34-41.

32 Ryder, M. L.; Stephenson, S. K. 1968: *Wool Growth*. Academic Press, London. 805 pp.

33 Smith, C. 1964: The use of specialised sire and dam lines in selection for meat production. *Anim. Prod. 6*: 337-344.

34 Spriggs, J. 1975: The economic implications of selection for twinning in Merino sheep. *Agric. Record 2* (3): 69-73.

35 Thatcher, L. P.; Napier, K. M. 1976: Economic evaluation of selecting sheep for wool production. *Anim. Prod. 22*: 261-274.

36 Turner, H. N. 1976: Methods of improving production in characters of importance. In *Sheep breeding*. (Eds G. J. Tomes, D. E. Robertson and R. J. Lightfoot). Western Australian Institute of Technology, Perth. 81-99.

37 Turner, H. N. 1977: Australian sheep breeding research. *Anim. Breed. Abstr. 45*: 9-31.

38 Turner, H. N.; Dunlop, A. A. 1974: Selection for wool production. *Proc. 1st World Congress. Genetics Applied to Livestock Production, Madrid, 1974. I*: 739-756.

39 Turner, H. N.; Young, S. S. Y. 1969: *Quantitative genetics in sheep breeding*. Macmillan, Melbourne, 332 pp.

40 Udy, M. 1976: Ram price formula based on records. *N.Z. Farmer 97* (19): 37-38.

41 Wickham, G. A. 1972: Some aspects of the relation of fibre fineness estimates. *Wool 5* (3): 32-37.

42 Young, S. S. Y.; Turner, H. N. 1965: Selection schemes for improving both reproductive rate and clean wool weight in the Australian Merino under field conditions. *Aust J. Agric. Res. 16*: 863-880.

7
Improving the Efficiency of Breeding Schemes

G. K. Hight
Deceased, Formerly Whatawhata Hill Country
Research Station, Hamilton
Consultants: A. L. Rae, J. N. Clarke, H. H. Meyer, A. G.
H. Parker

SUMMARY

This chapter examines the implications of a number of techniques that could assist in the genetic improvement of the sheep population. The use of group breeding schemes where the ram-producing nucleus flock is improved by the introduction of high-producing ewes from 'contributing' flocks, which in turn use rams bred in the nucleus, is examined. Evidence of their effectiveness and reasons for their effectiveness in comparison with the traditional 'closed-nucleus' system are presented. Methods of establishing flocks to ensure satisfactory gene-pools are discussed. Advantages and disadvantages of using ram lambs and mating ewe hoggets in order to increase the rate of progress through shortening the generation interval are compared. The use of artificial insemination is discussed but the technique is not recommended at present and it is suggested that natural mating with high ram : ewe ratios is better from cost : effectiveness considerations in most instances. Embryo transfer will probably be restricted in use to research studies and the multiplication of rare genetic material.

INTRODUCTION

To make rapid increases in genetic merit on an industry basis, large numbers of rams of high breeding value are required. Approximately 900,000 rams are used each year in the New Zealand sheep industry[18] and even with more extended use of sires there is obviously considerable scope for extending recording and improving the efficiency of breeding rams.

The rate of genetic progress is proportional to the difference in breeding value of the sires used and the average value of the flock in which they are joined. It is therefore particularly important that this difference be as large as possible.

In the *traditional breeding structure* flocks registered with the breed societies supply sires for the industry. Within each breed there tends to be a *pyramidal structure* with a small number of flocks at the apex supplying most of the registered sires to an intermediate but larger stratum of flocks which supply most of the rams used in non-registered flocks.

In this simple structure the genetic level of the commercial flocks depends on the genetic level of the registered flocks. This is an efficient system of genetically improving a breed, provided the flocks at the apex are in fact genetically superior to the intermediate units, which are in turn genetically superior to those in the commercial sector of the breed. If this proviso does not hold, then the system leads to no improvement, or worse, counteracts the improvement made by ewe selection within the commercial flocks. Unless the studmasters at the apex of the pyramid are effective in rapidly improving their flocks it is unlikely that there will be a significant difference in the genetic merit of flocks at the various strata of the pyramid because the pyramidal structure is an effective means of reducing genetic differences between strata in the breed.

The problem of improving the efficiency of breeding schemes so that larger numbers of rams of greater genetic merit are available then crystallizes to either:

(1) Ensuring that the maximum genetic gain is made in flocks in the top strata and that these gains are passed on as efficiently as possible to lower strata; or

(2) Eliminating the stratification where stud and commercial sheep are distinguished by registration based on pedigree so that rams are bred from both stud and commerical sheep which have proven recorded genetic excellence.

The costs of recording preclude objective measurement of performance in all sheep flocks, so the emphasis must be on getting maximum gain in those flocks supplying sires to other ram breeding units, whether the strata remain mainly based on registered sheep or not. Many aspects of this have been detailed in earlier discussions of selection and recording.

GROUP BREEDING SCHEMES

In *group breeding schemes,* high producing animals from several flocks are continually identified for transfer to a common *nucleus* unit in which sires are bred for transfer back to *contributing* flocks. The integration of these activities is the primary function of breeding groups and is outlined in Fig 7.1. Much larger numbers of elite animals can be made available and the continued superiority of the nucleus above the average of the contributing units can be assured. Breeders in many groups do not restrict their search for nucleus stock to registered animals but can select on performance from commercial as well as stud stock. In the traditional system these animals would not be used for ram breeding.

There are three potential disadvantages to such an open-breeding programme. First, the introduction of any animal increases the risk of

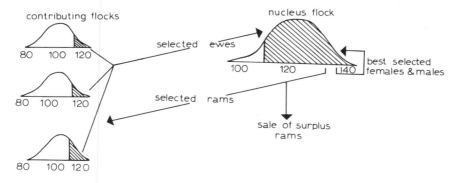

Fig. 7.1: An illustration of the distribution of performance levels of sheep in contributing flocks and how screening a small proportion of the highest performing ewes (hatched area) from each contributing flock results in an overall high level in the ram-producing nucleus.

introducing disease. Although this does not seem of major importance, continued vigilance on animal health aspects is essential. The second is the difficulty of making valid genetic comparisons among animals from different flocks because of the different environmental conditions in which these perform. By assuming that the average genetic merit of animals in different flocks of a scheme are similar and selecting a small proportion of animals that are superior to the average performance of each flock it is anticipated that this problem can largely be overcome. The extent that this assumption does not apply will affect the accuracy of selective screening. A third problem is that the traits used for screening commercial ewes from contributing flocks must be easily assessed or costs will increase markedly.

Relationship of Nucleus and Dependent Flocks

At present some group breeders are selecting from commercial flocks only, others from registered flocks, and still others from both registered and commercial flocks.[30] In two-tiered structures where contributors draw sires directly from the nucleus there is a minimum time lag in the transfer of genetic material. When screening is from registered sheep only, the nucleus usually supplies rams to other stud breeders who in turn produce rams for sale to commercial producers in much the same way as the traditional stud structure. This is then a three-tiered structure and the time lag for improvement to be transmitted to the commercial flocks is approximately doubled compared with the two-tier system.

The relationship between the nucleus and dependent flocks drawing rams from it in a two-tier structure is illustrated in Fig 7.2.

When first established the nucleus tends to have sheep with a considerably better average genotype than those in the contributing flocks. Transfer of genes from the nucleus to the contributing flocks, via

171

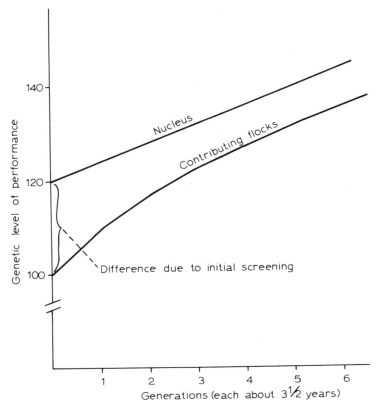

Fig. 7.2: Graph illustrating the differences in genetic level of performance between a nucleus and its contributing flocks.

the rams supplied, lifts the average genotype of the contributing flocks towards that of the nucleus. After the genetic level of productivity in the contributing flocks reaches a point that is below the nucleus by an amount equivalent to two generations (about seven years) of genetic gain in the nucleus, both sectors of the scheme make progress at the same rate. The time taken to reach the same rate of gain depends on the difference in genetic level at the start. Thus improvement is accumulated in each tier of the pyramid breeding structure at the same rate: However, each step in the multiplication of nucleus sires results in an 'improvement lag' approximating 2 generations.[3] This will be reduced if the contributors receive rams that are better than the nucleus flock average.

Screening of Sheep

The criteria used for screening females to the nucleus need to be related to the selection objectives of the group breeding scheme and the costs of

172

recording these criteria. Cost considerations can be very important as large numbers may be screened but few animals will be used for ram breeding.

For any given breeding objective it is theoretically possible to define the effect of screening a given number of sheep from a population of increasing size on the superiority of a nucleus flock compared with the average of the population from which these are selected. An example will illustrate this principle.

Assume there is a population of 100,000 ewes requiring 833 new rams for mating each year (1:60 ram to ewe ratio, 2 year mating 'life'), and that these rams can be bred from about 3,700 ewes (50:50 sex ratio, 80% survival to selection age and 60% of males bred used as sires). The ewes could be selected from increasing proportions of the total ewe population. The effect of this on the initial *phenotypic superiority* of the nucleus for number of lambs weaned only is given by curve P$_{NLW}$ in Fig 7.3, obtained by multiplying the standardised selection differential (i) by the phenotypic standard deviation (σ_p) in this case 0.65. The *genetic economic superiority*, in cents, for number of lambs weaned in curve G$_{NLW}$ is obtained by multiplying the values used for curve P$_{NLW}$ × heritability × relative economic value (Rev) = P$_{NLW}$ × 0.10 × 554. Similarly, curve P$_{GFW}$ is the expected initial phenotypic superiority of nucleus ewes selected only on hogget fleece weight (σ_p = 0.45), and G$_{GFW}$ the genetic economic superiority (h^2 = 0.30, Rev = 92 cents) resulting from screening ewes from populations of increasing size.

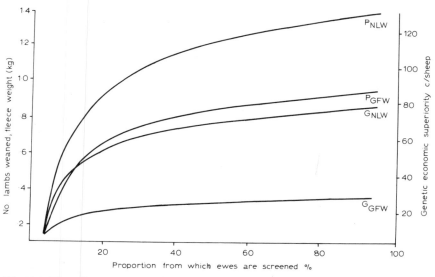

Fig. 7.3: The effect of screening from an increasing proportion of a total population of 100,000 ewes on phenotypic and genetic superiority of the nucleus ewes for number of lambs weaned (P$_{NLW}$, G$_{NLW}$) and fleece weight (P$_{GFW}$, G$_{GFW}$).

The initial level of genetic superiority of the nucleus above the average of the population from which these animals are selected will depend on the proportion selected as well as the heritability, variability (phenotypic standard deviation) and relative economic value of the traits considered.

For number of lambs weaned the phenotypic superiority (PNLW) of the nucleus from screening increases rapidly at first with increases in the population size screened, and then the rate of increase in superiority declines after about 20 times the number of nucleus ewes required are being screened. The response to increases in the population size screened is less with the more highly heritable trait, fleece weight. Thus for fleece weight a population about 5 to 10 times the size of the nucleus would be about optimal for that trait.

The genetic benefits of screening can thus be estimated for any trait or combination of traits. These genetic benefits must be greater than the costs involved. In practice more sheep are usually screened than are theoretically required for the nucleus, since sheep with structural and other obvious defects should not be transferred from contributors' flocks to the nucleus. The screening process provides the opportunity to establish at the outset a nucleus flock of high merit. It would take several generations to reach on the same level when relying solely on selection within a smaller closed flock.

In group breeding schemes involving the breeding-ewe breeds (such as the Romney, Coopworth and Perendale), the trait on which initial screening is based is usually number of lambs weaned by the ewe. A common requirement is that ewes eligible for transfer must rear twins after an unassisted lambing at two years of age. In some intensively-recorded groups, ewes are required to have lambed as hoggets and produced twins at their two subsequent lambings. In other cases hoggets of high body weight that have exhibited oestrus and are free from structural defects are selected. The screening may also involve a two-stage process with lambing hoggets or twinning two-year-old ewes being retained within contributors' flocks, and recorded through Sheeplan before selection of nucleus replacements is undertaken. One of the major reasons for the successful adoption of group-breeding schemes in New Zealand has been the simplicity of the criteria used for screening but the genetic efficiency of the screening varies considerably between groups.

Fixed standards of performance may not be desirable. Rather, the highest performing stock from each contributor should be identified each year. Contributors' flocks perform under widely different environmental conditions and arbitrary phenotypic standards for entry may exclude animals of acceptable genetic merit. Thus superiority should be related to each flock average.

The relative emphasis that should be given in screening for fertility, hogget body weight or other traits is indicated in Chapter 6 and by the weighting factors used in the Sheeplan selection index. Where indirect measures of performance[23] for example number of hogget oestruses, are used as the criteria for screening then the heritability and genetic correlations with the primary trait (fertility) or breeding objective need to be defined. Finally the feasibility of recording one or more traits and the costs involved must be evaluated before group members decide on the trait(s) that are most suitable for their particular objective and circumstances. These aspects are critical to the economic efficiency of any breeding programme.

Sheep breeders usually have definite ideas about visible sheep characteristics which could be related to the animals' ability to thrive and produce well and to commercial viability. These characteristics include structurally sound feet and pasterns, absence of scouring, general appearance and thriftiness of the animal and face-cover. Unfortunately the relationship of these characteristics to performance in high-producing animals is not clearly defined. The argument can be advanced[32] that animals with high production are those which are capable of producing well in a given environment, so that measured production itself should be the criterion for ability to thrive and produce. The level of production at one point in time may not however allow for costs of production over the life of the animal in a given environment and management system (e.g. for footrot treatment, crutching of lambs for flystrike) so that *if* these factors are heritable they may have high economic merit. The maintenance of a balance between artificial and natural selection may be desirable to ensure a continued high level of phenotypic performance in several environments.

The approach usually adopted is to record and select for high production of measurable heritable traits of known economic importance in an environment in which the offspring of selected animals are expected to perform. The "easy-care" traits are then considered amongst those sheep of highest genetic merit. As further information on nutrition, and physiological genetics (e.g. facial eczema resistance) enables an adequate definition of these 'ability to thrive' traits, then it will be possible to use this information in an objective breeding programme. Until this is done the emphasis given to 'easy-care' traits remains subjective and a matter of some conjecture. In large group breeding schemes there are many animals of high and often similar breeding value. In the selection of ewes or of nucleus sires it is often possible to apply selective culling for 'ability to thrive' or easy-care traits without markedly influencing the superiority of the selected group for measured productive traits.

At this stage only two relatively small co-operative breeding schemes have been started with the prime lamb sire breeds in New Zealand. The same principles can be applied to these and other breeds as is now applied to breeding-ewe flocks in existing schemes. Differences in breed structure, population size of commercial flocks, selection objectives and other factors may however, require a different emphasis in breeding methods to identify genetic merit.[30] Non-registered straightbreds will not normally be available in large numbers for the meat breeds.

Selection Within the Nucleus

Central to the efficient operation of breeding schemes is the management of stock, and the objective recording of performance data.

The mechanics of actually undertaking the selection of all classes of stock need to be clearly set out to maintain consistency and efficiency. This is particularly important in the selection of sires. Updated estimates of the breeding values of all classes of stock need to be available in a condensed, easily accessible form for breeders to use in the yards or woolshed. Computers have made this objective readily attainable, and are indispensable for the operation of an efficient breeding scheme.

The selection of sires should commence with a careful (paper) inspection of the selection index values, and of component traits of all sires available. Those rams of sufficient merit for consideration as nucleus sires should then be cut-out from the flock and formed into selection groups on the basis of the selection indices. Starting with the top ranking groups, each potential sire needs to be carefully examined for absence of horns, structural defects of the teeth, feet, pasterns and sexual organs, black fibres or other functional defects of economic significance, until the required number of nucleus sires is obtained. In group schemes it is preferable that a panel of three or more members should select both the nucleus sires, and those rams to be made available for selection by contributors for use in their own flocks. Every care is needed to avoid human error through careful checking of tag numbers, performance records and by systematically working through a list of selection points. Selection should be undertaken in sheltered well lit yards, on clean level concrete and with all records readily available. Copies of the records applicable to the sheep being selected should be available for each member of a selection panel. A systematic approach sets objective standards for all members to follow.

Comparison of Gains from Open and Closed Nucleus System

If the sire-breeding central nucleus is continually open to gene inflow from other levels of the group structure there are two main effects. First, the genetic gains from selection in contributors' flocks are introduced to the nucleus, and can be passed from the nucleus to the whole population. Secondly, by the regular introduction of less closely related individuals to the nucleus, the rate of inbreeding is reduced.

The theoretical rates of genetic gain which can be obtained by opening the nucleus to females from contributors' flocks have been studied.[20] [22] [31] An increased rate of genetic gain is expected in an open compared with a closed nucleus, mainly because of the greater selection intensity possible in females. With a trait such as fleece weight, measurable in both sexes, the open nucleus may give genetic gains 10-15% greater than in a closed nucleus. Using a trait which can be measured only in the ewe (such as weight of lambs weaned per ewe, or number of lambs weaned) then the rate of gain over the closed nucleus may be in the range of 15-25%.

The optimum group breeding structure is likely to have about 20 times the number of sheep in the nucleus being screened, 5 to 10% of the population in the nucleus,[22] [31] about half of the nucleus female replacements introduced from outside, and all surplus females born in the nucleus used for breeding in contributors flocks. However, the rate of genetic gain is not very sensitive to changes in the proportion of nucleus female replacements selected from contributors' flocks.

The open-nucleus may be particularly valuable for improvement in all-or-none traits, such as fertility. The effective heritability of fertility could increase as fertility level increases, and with intensive screening and selection of replacements the fertility level may be raised enough to make a significant improvement on the accuracy of selection, rate of genetic improvement and selection limits.

The introduction of breeding animals from contributors' flocks to the nucleus reduces the average relationship between nucleus sires and dams below what it would be in a closed nucleus of the same size. An open nucleus in which half the females are introduced from contributors has about twice the effective size of a closed nucleus of the same number of sheep. The rate of inbreeding should then be approximately halved. Despite this, systematic avoidance of mating closely related sires in the nucleus should be practised because some systems tend to select sires closely related to a small group of nucleus parents.

Thus it is theoretically possible to screen a large population for sheep of high productive merit, to form a foundation nucleus flock, and by continued screening and selection to maintain the genetic superiority of the nucleus ram-breeding unit above that of the contributor flocks. The higher this difference in breeding value the greater the average rate of improvement that will result from the use of nucleus-bred rams in contributors' flocks. The maximum theoretical advantage is however, unlikely to be fully realised in practice because of the lack of complete records on all animals in the screened population, or the use of indirect criteria of merit.

Performance of Nucleus Flocks

The extent to which the screening and initial selective breeding has made it possible to quickly establish large flocks of high performance relative to the population from which these were screened is indicated by the phenotypic levels being achieved in nucleus flocks of group schemes.[19][29] The lifetime fertility level of the nucleus ewes in the Lands and Survey Waihora Romney breeding scheme[19] at mating in 1977 is given in Table 7.1. These data include hogget lambing records for all Waihora bred replacements but not the hogget record for twinning ewes screened in after their two-year-old lambing. The number of ewes with different lifetime breeding values for number of lambs born per number of lambings for all Waihora and Otutira ewes is given in Fig 7.4. There

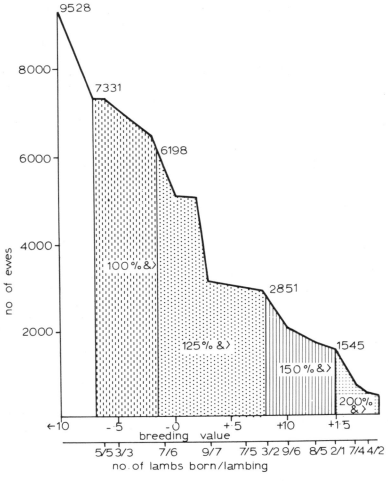

Fig. 7.4: Distribution of lifetime fertility of ewes in the Waihora flock in 1977.

TABLE 7.1: Lifetime Fertility of Ewes Present in the Waihora Flock
1977

Age in 1976 (years)	No. ewes	% lambs born /lambings	% lamb survival rate	% lambs weaned /lambings
1	1744	66.4	83.1	55.2
2	903	127.1	90.9	115.6
3	778	138.2	92.3	127.5
4	575	156.4	91.3	142.7
5	231	144.5	91.3	131.9
6	246	159.8	90.7	145.0
7	145	162.8	89.1	145.0
8	60	174.6	89.4	156.2
9	26	166.0	89.8	149.0
2 to 9-yr-old	4708	147.8	89.0	131.5

Hogget lambings included except for ewes screened to nucleus

were 1,545 ewes with a lifetime fertility equivalent to 200%, and 2,851 ewes with a lifetime fertility equivalent or higher than 150% in this recorded nucleus. Clearly, large numbers of high fertility ewes are present, illustrating that it is practicable to quickly form and maintain a large elite nucleus flock.

There is of course considerable genetic variability and opportunity for further selection among the animals present in a nucleus. There was a range in body weights of from 22 to 49 kg and in fleece weights from 1.8 to over 4.1 kg in the ewe hoggets in the Waihora flock in 1975.

Evidence of Genetic Progress

Direct scientific evidence that effective genetic progress is being achieved by group breeding schemes or indeed other commercial selective breeding programmes is limited. No private group schemes or studs have maintained controls to evaluate genetic progress as these are orientated to breeding rams for use within the group or for sale.

The one direct sheep comparison for which information has been published comes from the study comparing the performance of the progeny of rams bred in the Waihora scheme with the progeny of rams purchased from private breeders. In this project a flock of ewes was divided at random into two groups. One group was mated to Waihora rams and the other to purchased sires, with the progeny subsequently being joined with Waihora or commercial rams respectively. Fertility data for this comparison are given in Table 7.2. In 1973, 705 two-year-old ewes sired by Waihora-bred rams weaned 22% more lambs than the 711 ewes sired by purchased 'commercial' rams. In

TABLE 7.2: Comparison of Progeny of Waihora (W) and "Commercially" bred (C) Romney sires

Year	Age of ewe progeny	assisted births (%) W	C	twin births (%) W	C	LD/EM (%) W	C
1973	2-year	7.8	14.1	27.4	10.0	101.7	79.8
1974	2-year	5.4	7.4	15.7	9.9	102.0	91.8
	3-year	8.1	8.2	19.2	8.9	105.0	98.0
1975	2-year	5.9	11.0	13.9	7.4	96.4	85.5
	3 & 4-year	3.6	9.3	37.3†	24.4*	122.1	108.3

† includes 6 sets of triplets;   * includes 2 sets of triplets

1974, a less favourable season, the difference for two-year-old ewes was 10.4% and 7.0% for three-year-olds. In 1975 two-year-olds weaned 10.9% more lambs, while the combined three-and four-year-old ewes weaned 13.8% more lambs. The increased weaning percentage was due partly to a higher multiple birth rate and to a better lamb survival rate. Other advantages in the rate of finishing of wether progeny for slaughter and live weights of ewe hoggets have been apparent. This initial superiority is substantial and probably represents almost entirely the gain achieved by screening rather than selection within the nucleus. When the results are compared with the gains in fertility from within-flock selection in long-lasting small research experiments[8] it suggests that screening for fertility has been successful. Further demonstration and evaluation of the advantages of screening and adoption of the group breeding scheme approach, compared with rates of genetic gain by objective selection within closed units and with subjective traditional methods, is required.

Selection from several populations seems likely to provide better foundation stock[21] with a greater degree of genetic variability, higher rate of genetic gain and a higher upper limit of selection response than selection from one flock. Larger total effective population size could be associated with slightly greater total response to selection. A method of further increasing the effective size of group breeding schemes and increasing genetic variability is to exchange rams between the nucleus units.

Business Organisation

There is a need for a formal business framework in the operation of group schemes. The close contact such an organisation can exert over all phases of ram production ensures members have a more cohesive, effective selection programme than if this was attempted by independent breeders. A sound business organisation may also be used to extend the field of operations in many spheres — such as advertising, marketing, application of specialist techniques, processing of animal products or any other activity conducted in the interests of group members.

The form of business organisation adopted to suit the breeding and financial objectives varies widely. These include private companies, Industrial and Provident Societies, partnerships, or where the nucleus operator owns the stock in a 'gentleman's agreement'.[25] The organisation most favoured is that of a company structure with contributors having equal shares and voting rights, and with the company owning all or part of the nucleus. This provides a workable flexible arrangement in which members have continued interest, access and incentive to assist the nucleus operator and to represent the interests of group members in various activities. The number of members can readily be expanded or contracted through the sale of shares. Some companies operate on the principle of nil profits from the nucleus with members contributing ewes to the nucleus in exchange for rams on a 5 or 4:1 basis, and also receiving other selection rights for nucleus sires on a roster system.

Effective communication is essential between people involved in any breeding organisation. To this end directors or members of group schemes need to meet at regular intervals and all members be kept informed of activities through newsletters, visits to properties, and by circulation of copies of nucleus flock records, or summaries of performance. Regular reviews of all aspects of co-operative breeding structures should be undertaken.

Reasons for the Success of Group Breeding Schemes

Several reasons can be advanced for the success of group breeding schemes. Some of these are not strictly genetic but rather relate to organisational and economic aspects[16] of their approach to sire production. They include:

(1)   The opportunity of rapidly establishing and maintaining a foundation nucleus flock of high productivity through the high selection intensities that are possible in both males and females.

(2)   The narrowing of breeding objectives towards those characteristics related to commercial economic merit through the need for co-operating members to reach a consensus of agreement. The nucleus flock operator and directors of group breeding schemes need to continually justify their selection and business decisions to other members of the group, thereby ensuring relevance of the breeding programme.

(3)   Acceptance of a uniform recording service such as provided by Sheeplan, and the centralized recording of nucleus animals in a common environment to avoid the biasses inherent in comparative recording.

181

(4)   A breeding group has a better identity, because of its size, than any individual breeder. This helps the group to be better known among farmers and helps market surplus rams.

(5)   Economies of scale and of integration can enable large numbers of sires of known breeding value to be produced and marketed at competitive prices.

(6)   Overhead costs are reduced as only a selected but large population is recorded in detail.

(7)   Breeders share their facilities, time and capital so that the input of each member is reduced. More effective use can also be made of superior sires within a large nucleus than in small units.

(8)   The co-operation and mutual discussion, the element of competition over traditional approaches, and the continued direct involvement in the breeding of sires stimulates those involved.

(9)   Because of their co-operative nature and size, group schemes act as a focus for technical personnel. These factors tend to result in the use of more sophisticated techniques.

(10) The structure gives the group the potential to extend its range of activities.

Cost-Effectiveness Considerations

In view of the widely differing costs and likely impact of alternative breeding schemes on the industry it is not sufficient to compare these solely in terms of genetic improvement. Their cost-effectiveness must also be considered.

In group breeding schemes where only high-performing unregistered ewes are identified for detailed recording a high proportion of their male offspring should be above the contributed flock average in breeding value. The purchase cost of rams sold to clients outside the group need not be high for such groups to remain commercially viable. This gives an immediate potential economic advantage over ram breeding flocks based on limited selection among more-expensive registered sheep. This is particularly true when a limited market for high priced individual rams exists in the industry. The economies of scale of operation, and the principles of integration can more readily be applied in large group schemes than within closed units.

The formation of large nucleus flocks has stimulated the development of facilities, and new approaches to recording and selection procedures so that larger numbers of stock can be recorded accurately and quickly with the minimum of labour. For example, numbered neck tags and larger ear tags have removed the necessity to catch lambing ewes to identify offspring by dam number. Better weighing equipment and data recording sheets have made it feasible to weigh over 300 sheep per hour, or to record the fleece weights and quality grades of 1,200 sheep/day.

182

Further technical developments are quite feasible.

However, few economic studies have been undertaken of the influence of breeding structure, scale of operation or other variables on the cost-effectiveness of applied ram breeding. To take account of the time scale involved, both in effecting improvement and realising the gains, returns have to be discounted to the year a scheme is started in order to allow for the loss of interest and risk element on invested capital. One study (G. B. Nicoll & C. A. Morris, pers. comm.) of the Waihora sheep breeding scheme showed a net cumulative return above opportunity farm costs of $5.52m. with a 10% annual discounting rate over 15 generations. In this programme every 1,500 additional rams produced per year were worth a cumulative total of $257,000 to the scheme as a whole. While further study is required it seems likely that group breeding schemes are more economically viable than where rams are bred from high-cost registered animals in small units with high labour costs.

National Nucleus Units

An extension of the concept of screening and selection of productive sheep to form group breeding schemes is to select sheep from the nucleus units of such schemes, and from other recorded flocks, to form one or more National Breeding Flocks of various breeds. With a high selection intensity and rapid generation turnover, sires of known high comparative breeding value could be produced from these national nucleus units for mating within contributing ram-breeding flocks. More effective dissemination of genetic material of high merit might then be achieved on a national basis. Such national nucleus flocks could be established under the direct control of a producer representative body — such as a Breed Society or the proposed National Sheep Improvement Council. It is important to the industry that the benefits of selective breeding should be passed on to the commercial producer as quickly as possible. Those ram breeders using outside sires need recognised sources of rams, or sons of outstanding sires which are genetically capable of improving performance.

Any ram breeder recording through Sheeplan (registered and non-registered and group breeding schemes) should be eligible to contribute ewes (and initially) rams of high genetic merit to these national flocks according to rules decided by a controlling body representative of the breeder users. Intensive selection among those nucleus sheep and their progeny could be expected to generate rams of high genetic merit which could be ranked in order of breeding value, for subsequent multiplication by ram breeders and thence for the industry. While the number of rams of superior merit that could be generated from breeder-controlled national flocks of say 500 ewes and their progeny would be small in relation to all rams sold between ram breeders, these rams could markedly influence the rate of genetic improvement of a breed.

By grazing the nucleus flocks of different breeds separately but in a defined common environment, valuable information on the relative levels of performance and efficiency of food conversion could be obtained. National nucleus units would also provide a better prospect of long-term continuity than in private flocks or group breeding schemes where the retirement of key participants can result in uncertainty about long-term stability. The initiative for the development of national nucleus units and their control must come from ram breeders themselves rather than an independent group.

ESTABLISHMENT OF GENE POOLS

One of the most important factors governing selection response is the genetic variation of the flock. In small closed breeding populations, intensively selected for a particular trait, it has been found that response to selection tends to decline after several generations. This problem may be overcome by increasing the genetic diversity of the group selected.[2]

The existence of a large degree of genetic variation within a breed is well established and it may in the shorter term be commercially more practicable to exploit this variation rather than the genetic differences between breeds, particularly when this involves the marketing of unfamiliar stock. However, the potential exists to widen the gene pool by crossbreeding and then selecting within the resultant crossbreds.[28]

Crossbreeding has played a part in the development of most sheep breeds, and a number of synthetic breeds have developed in New Zealand from inter-breeding and selecting crossbreds (see Chapter 2). In most cases the synthetics are based on two parent breeds but more recently there has been a tendency to include more breeds in the gene pool.

The usual justification for the establishment of these synthetic breeds is that there is no single breed with sufficient variation or desired combination of traits required. This leads to the crossing of breeds each of which have a satisfactory level of performance in one or more of the desired characteristics. When the first cross proves superior to the existing straightbreds it is then common to interbreed the crossbred to obviate the need to maintain separate flocks of the parent breeds. The Coopworth and Perendale breeds were developed in this way. While the the loss of hybrid vigour in reproductive and maternal traits associated with interbreeding usually results in the interbreds being inferior to the first crosses (see Chapter 5), intensive selection may compensate for the loss in heterosis.

It has been suggested that, even if a crossbred base flock is not initially quite as good as a straightbred flock, the presence of a wider array of genes may enable a greater response to selection and hence the crossbreds to evenutally surpass the straightbreds.[32] However there is little experimental evidence to support this contention[24] and the returns from the sale of rams in the early stages of such a venture would be unlikely to offset the costs of recording in the early years.

One situation where crossbreeding may be warranted is when a combination of high fleece weight and wool fineness are selection objectives. This combination requires an increase in follicle density. The Merino appears to have genes for follicle density which are very rare in British breeds. Thus in Perendales where fineness combined with fleece weight is a commonly stated objective, an infusion of Merino genes might allow for a greater response to selection.

The relative merits and costs of selection within breeds and cross-breeding need to be defined before breeders embark on crossbreeding programmes or the establishment of new synthetic breeds.

INTRODUCTION OF EXOTIC GENOTYPES

Genetic variability for certain characteristics, limited in the local population, can sometimes be supplemented by the importation of exotic breeds.[5] [32]

Exotic breeds could be used as purebreds to replace existing genotypes, to cross with local breeds or to develop new synthetic populations. The establishment of new breeds might also be highly profitable in view of their possible re-export potential. Unfortunately animal health safeguards make this an expensive process with a long delay before the descendants of the imported stock become available to the industry assuming no exotic diseases appear. The only country from which sheep can usually be imported into New Zealand without stringent quarantine is Australia and regulations in importations from different regions of Australia are subject to change according to the disease situation. Since most Australian sheep breeds are already well represented in New Zealand the facility to import from Australia does not result in a much greater gene pool being available. From all other countries sheep can only enter New Zealand through quarantine establishments.

The choice of breeds for importation requires a clear definition of breeding objectives and the testing of the genetic merit of production traits for exotic compared with local breeds under local conditions. Ideally, initial experimental evaluation needs to be followed by on-farm testing and demonstration in several environments. All of this involves considerable time and expense.

Although new genotypes may be found to have major advantages they are also likely to display disadvantages of one kind or another. Effective exploitation of new genotypes may require considerable modification of farming and product-handling systems.

The existing investment of breeders in their current stock is very considerable so that a change of breeds can involve considerable redeployment of capital, and this may be difficult when existing investment returns are low. Consultation is therefore needed with industry on an equitable method of releasing any useful exotic livestock or their crosses. The likely return on investment for individual breeders and industry from increases in genetic potential through within-breed selection, crossbreeding, utilisation of exotic breeds or other methods requires further economic analysis.

GENETIC RESERVOIRS

The formation of national genetic reservoir flocks may be warranted to ensure:
(1) that exotic sheep imported and tested at considerable expense are maintained even though they are not used immediately;
(2) that other types of sheep in danger of extinction but which have features of potential importance are maintained;
(3) the generation of new strains of sheep with features which may not be immediately important but which have potential in future markets.

To ensure reasonably effective selection among sheep represented in such genetic reservoirs and valid comparisons of productive performance among them of derived synthetics each 'breed' should be represented by at least 150 mature females in one location.

The development of genetic reservoirs could provide invaluable flexibility to meet new market opportunities and for development of new synthetic breeds. This would be a risky commercial undertaking for private breeders and therefore technical and industry support would be needed.

REDUCING THE GENERATION INTERVAL IN PRACTICAL SHEEP IMPROVEMENT

At present in most ram breeding flocks ewes and rams are mated first as two-tooths. In Chapter 4, it has been mentioned that at the same intensity of selection, faster genetic progress can be made if the generation interval is reduced. It may be feasible to reduce the generation interval and to increase efficiency by mating both ewes and rams in their first autumn, and retaining older ewes and rams for limited numbers of matings. Unfortunately most methods of reducing the generation interval tend to reduce the accuracy of predictions of the breeding value.

Mating of Ram Hoggets

The first problem in the use of ram hoggets as sires is the availability of sufficient information on their dam's fertility for accurate selection in fertility to be undertaken. Usually only a proportion of their dams will have had two or more recorded lambings. Other traits such as body weight, fleece weight, wool quality, structural soundness of the teeth, feet or testicles may not be sufficiently developed in the young animal by the first autumn for reliable indices of overall breeding merit to be formulated. In Sheeplan the percentage reduction in the rate of improvement in the aggregate breeding value which would result if one of the traits were excluded from the selection index for dual-purpose breeds are respectively 24%, 2% and 13% for number of lambs weaned, weaning weight and hogget body weight. Thus by selecting ram hoggets as sires instead of two-tooths the accuracy in identifying those of highest breeding value could decrease by up to 13%. A further restriction is that some ram hoggets (e.g. twins born to ewe hoggets, or quadruplets) may not be sufficiently developed to have attained puberty.[14] However there may be a tendency for ram progeny of highly fertile parents to exhibit precocious sexual development.[4] By selecting those lambs likely to be of 'nucleus sire class' at weaning, and grazing these preferentially until final selection pre-mating it is possible to get a high proportion of such ram lambs well developed by their first autumn.

Breeders tend to select for mating those ram hoggets for which fertility genotypes appear outstanding, and from this group select sires which exhibit early apparent superiority in other traits. Further study of the efficiency of selection of sires at the hogget instead of at the two-tooth stage, in relation to their adult breeding value, the performance level of progeny and the influence on current ewe flock performance, are required before guidelines can be given as to the proportion of rams of different ages to use in a nucleus breeding programme.

Nevertheless some use of ram hoggets, to mate with ewe hoggets, two-tooth ewes, a proportion of the elite ewes in a nucleus flock, as 'tailing-up' rams to extend the ewe to ram ratios of older sires, as a method of obtaining access to genetic material that would not normally be available to the flock owner, or for progeny testing, may be justified.

In some group breeding schemes contributors use nucleus ram hoggets for a 30 day mating period before returning them to the nucleus unit.

Limited information on the mating efficiency of ram hoggets[26] indicates that they may be slightly less effective at detecting and mating oestrous ewes than two-tooth rams, particularly if several ewes are in oestrus at one time. This factor may decrease conception rates. However, ram hogget to ewe ratios between 1:30 and 1:100 have been used successfully with a restricted mating period of 35 days or less at Waihora. Careful selection (for fertility and breeding background, functional defects), preparation (through the provision of adequate nutrition), and attention to animal health aspects are clearly also of importance in the successful use of ram hoggets.

Mating of Ewe Hoggets

Studies of the occurrence of mating activity in ewe hoggets joined with harnessed teaser (vasectomized) rams have shown oestrous activity to extend from the end of March until late July, with a peak of activity from mid-May to mid-June.[6] [17] There is considerable variation in the proportion of ewe hoggets exhibiting oestrus both within and between breeds, and in different seasons. In Romney and Border Leicester × Romney hoggets[17] the mean weight at first oestrus was 34 kg for single-born and 31 kg for twin-born hoggets. The proportion showing oestrus increased with mean April live weight. The number of ewe hoggets exhibiting oestrus within a flock may also be influenced by their birth-rearing rank, dam-age and age during the hogget mating season. The proportion of hoggets within these sub-groups will then influence the proportion of hoggets mated, primarily through a live weight effect. For example, a smaller proportion of hoggets born to one-year-old dams will exhibit oestrus in comparison with singles born to older ewes.

The optimum nutritional level, and feeding system that is required for hoggets to attain maximum sexual activity and for optimum hogget and subsequent ewe fertility is still uncertain. It appears desirable to aim for well-grown hoggets (>35 kg) so that these will exhibit oestrus early in the breeding season, and which subsequently attain adequate (>50 kg) pre-tupping ewe live weights.

Factors favouring the mating of ewe hoggets in ram-breeding flocks include:

(1)   the early recognition of fertility potential[4] [7] and the provision of fertility data on the progeny of tested rams one year earlier than normal;

(2)   increased numbers of lambs weaned per female wintered, and more offspring available for selection as replacements or sale;

(3)   a lower age at first breeding resulting in a higher generation turnover and faster rate of genetic improvement if the hoggets that lamb are of superior genotype to the rest of the ewe flock.

Disadvantages include:

(1)   a longer lambing period as a consequence of the later mating of ewe hoggets;

(2)   a need to supply extra feed for hoggets before and after lambing;

(3)   more small lambs. Ewe and ram offspring may suffer a handicap in live weight which can persist to the two-tooth stage.

The magnitude of these disadvantages will be influenced by the level of grazing management, stockmanship, nutrition and genetic potential of the animals.

The breeding of sheep which have the inherent potential to rear offspring without assistance as ewe hoggets could markedly increase the efficiency of utilisation of seasonal feed resources per unit of female stock wintered, and profit. Direct measurement appears the best method of identifying genetic variation in this trait, rather than indirectly through genetically correlated traits, for example hogget oestruses, hogget body weight or ewe fertility.

In general, the literature [1] [11] [13] on mating and lambing ewe hoggets indicates the following:

(1)   The higher the live weight and later the age at mating during the autumn season, the better the hogget lambing performance both in terms of ewes lambing and number born.

(2)   Higher lamb losses may occur compared with mixed-age ewes. In 1976 lamb survival rates to weaning were 74.3% in lambs born to hoggets at Waihora compared to 83.6% in mixed-age ewes.

(3)   The mothering ability of ewe hoggets is generally very satisfactory as compared to adult ewes.

(4)   Live weight gains of single lambs reared by hoggets compare favourably with twins reared by two-year and older ewes. However since they are born later they tend to be lighter at a given time. This live weight difference decreases with age so that lambs born to hoggets may only be 2-3 kg lighter on average than those born to older ewes by 14 months of age if nutrition is adequate.

(5)   Breeding from ewe hoggets can check their growth. Until about 120 days of gestation the additional demands of pregnancy do not appear to have a marked retarding effect on growth, but lactation can reduce body weight and wool production. The time required for compensation of this difference in body weight between lambed and unlambed hoggets will vary with nutritional level, and time of weaning.

(6)   There is a possibility that the eruption of incisor teeth may be delayed in suckling ewe hoggets but, if this effect is present, it does not appear permanent.

(7)   The early breeding of well-managed and adequately-nourished ewe hoggets has little or no detrimental effect on subsequent reproductive efficiency, and there is some evidence of improved fertility, lower dystokia at the second parity and of enhanced mothering ability at subsequent lambings.

The level of nutrition, management and genotype will therefore largely determine the feasibility and contribution to genetic improvement of mating ewe hoggets. Experience in the Waihora breeding scheme may act as a guide for the adoption of ewe hogget lambing in other ram breeding flocks. Ewe lambs were preferentially fed as well as possible on pastures with a high white clover content from weaning, with those born to one-year-old dams usually being grazed separately until mating. All were shorn in April. The mating period was restricted to about 30 days from late April to avoid small late-born lambs at weaning. Both ram hoggets and/or two-tooth rams at ewe to ram ratios varying from 1:60 to 1:100 have been used successfully. At crutching immediately before lambing those hoggets in-lamb were identified by udder development and grazed separately. A minimum level of shepherding at lambing other than for progeny identification was practised. Lambs were weaned at about 60 days of age and the hoggets then shorn and fed to recover live weight. The aim is for a concentrated lambing and early weaning to increase the time available for recovery of live weight before the two-tooth mating.

The average level of fertility in the Waihora ewe hoggets between 1970 and 1976 is given in Table 7.3. The data are not strictly comparable because of varying proportions of hoggets born to one-year-old dams in different years, but hogget lambing has increased the number of lambs weaned on a flock basis by about 20%.

TABLE 7.3. Ewe Hogget Lambing Performance, Waihora

	Year						
	1970	1971	1972	1973	1974	1975	1976
No. hoggets present	312	573	839	1632	1922	2162	2665
% hoggets lambing	63.8	61.6	52.0	51.1	49.2	64.4	51.3
% lambing twins	2.0	4.0	5.5	3.5	5.6	3.3	4.9
% lamb survival rate:							
singles	66.7	66.4	83.3	82.0	78.5	79.3	76.4
twins	50.0	50.0	72.9	69.0	58.5	53.3	53.7
Total	66.0	65.1	82.4	81.1	76.4	77.6	74.3
% lambs weaned/hoggets lambing	67.3	67.7	86.7	83.9	80.6	80.2	77.9
% lambs weaned/hoggets present	42.9	41.7	45.1	42.9	39.6	51.7	40.0

In summary, limited comparative information suggests that the mating of ewe hoggets can result in higher profitability, increased numbers of ewe replacements, early identification of fertility potential, and the generation of additional rams for selection as nucleus sires or for sale. Special attention does however need to be given to nutritional and other management factors, and these aspects may restrict the general practice of hogget lambing. Further information on the effect of hogget lambing on the production of ewe hoggets from different ages of ewes, and fertility genotype is required before the value of this practice on the rate of genetic progress can be more clearly defined.

ARTIFICIAL INSEMINATION

The technique of artificial insemination (AI) in sheep is discussed in Chapter 10. The obvious genetic advantage of AI is that a more widespread use of superior sires may be achieved than under natural mating, permitting more intensive selection to be practised among sires. Russian and Australian experience indicates that 1,000 to 1,500 ewes can be inseminated using undiluted fresh semen and about 2,500 ewes per ram using diluted fresh semen.[9] If the technical problems of storing ram semen can be overcome then large numbers of ewes might be covered per ram, and the mechanics of inseminating large numbers of ewes in different locations made easier and more economic. However, one theoretical study[12] indicated that, with intensive use of a small number of rams selected for clean wool production in flocks up to 1,000 ewes, the extra rate of improvement through use of AI over natural mating may decline year by year due to inbreeding effects on fertility and other traits. As flock size increases above 1,000 ewes the rate of genetic improvement in fleece weight under AI may increase sharply relative to natural mating and give a 20% to 30% increase in flocks of 5,000 ewes or more. This latter situation could be approached in large breeding schemes. In order to take advantage of intensive male selection through AI, a self-contained breeding group would need to ensure that its nucleus ram-breeding flock used a sufficient number of unrelated rams to avoid a decline in breeding merit from inbreeding.

The accuracy of assessing the breeding value of individual rams for fertility traits on the basis of their dam's fertility, or that of female relatives is not high, so that extensive use of rams through AI would make it necessary to progeny test potential sires before their final selection. The group breeding scheme structure with its contributor flocks, and the capacity to spread the costs of progeny testing sires over large numbers of sheep offers the opportunity to effectively use progeny testing. This may be particularly useful for traits of low heritability but high economic value (e.g. fertility), traits expressed in one sex, or of high commercial importance which may not otherwise be tested for (e.g. structural soundness of the feet, facial eczema resistance, inheritance of black fibres). Several groups are undertaking progeny testing trials and others are examining the possibilities of progeny testing, which is discussed further in Chapter 4.

A careful balancing of the selection accuracy, selection intensity, generation interval and inbreeding effects is necessary to design a breeding system using progeny testing and AI. The question of how best to utilise a proven sire once these have been identified is still open to considerable ingenuity. A clearer definition of the technical, genetic and management aspects of large-scale AI[15] is required before this technique could be recommended as a means of extending the use of proven sires in breeding programmes in New Zealand above that which can be obtained by extending ewe to ram ratios under natural mating.

EWE-TO-RAM RATIOS

If rams are mated to large numbers of ewes this reduces the costs and the number of rams required for mating, and also increases the selection differential. It has similar, although less intensive, effects to artificial insemination. If very wide ewe-to-ram ratios are used this can also increase inbreeding and the risk of undesirable recessive defects while decreasing genetic variability. The cost of extended use of rams through natural mating is much less than that of AI. However, the differences in estimated breeding values of potential sires are usually not large as the selection indices are also usually based on incomplete lifetime records of dam's fertility. In order to increase genetic variability, reduce inbreeding, and decrease the risk of defects in rams or their offspring being expressed at later stages, most flocks tend not to maximise the number of ewes joined per ram. Occasionally sires which appear to be of outstanding breeding value are mated with less than the maximum number of ewes as hoggets and/or as two-tooths, and then extensively (> 300 ewes) as four-tooths if progeny test data indicate this is justified. Such restrictions need not apply in commercial flocks (see Chapter 10). The age, number and genetic standard of the rams available and flock size are all factors influencing the optimum ewe to ram ratios in ram breeding flocks.

OVA TRANSFER

The technique of ova transfer as outlined in Chapter 10 was developed with the objective of producing greater numbers of offspring from superior females, particularly when combined with induced super-ovulation.

Technical problems[27] and associated costs have limited the application of the technique. The high value of exotic breeds of cattle has resulted in some commercial use in cattle breeding but there has been little exploitation of the technique in sheep flocks apart from in research projects.

In addition to these technical problems, the recipient ewes do not have the opportunity to express their own breeding potential for reproductive traits in that year,[9] and the donor ewes may have a biased performance record and delayed lambing as a result of the treatment and surgery. This delayed lambing could also have carryover effects on the age and sexual maturity of their offspring. The potential of egg transfer is further limited by the small contribution to genetic progress from selection on the female side. Even if the egg transfer technique is further refined it appears unlikely to have a general role in sheep breeding because of the costs involved, and its main application will probably be restricted to the multiplication of rare genetic material, and as a technique for research studies.

THE FUTURE ROLE OF BREEDS AND BREED SOCIETIES

Co-operative group-breeding schemes are now well established in New Zealand. The number of these schemes, using both registered and/or unregistered sheep is likely to increase and to be extended to other breeds. Increased numbers of performance-backed rams not eligible for registration with existing Breed Societies can be expected to be sold to commercial breeders.

Nevertheless, it is not expected that group-breeding schemes will take over the whole of the ram breeding function of different breeds. Rather group schemes will complement the traditional stud breeder and others. This will also create more competition and encourage efficiency of genetic improvement.

A breed is a genetically distinct group of animals, which is usually recognised as such by the establishment of a Breed Society. A breed society's traditional function has been to look after the affairs of its breed, including:

(1) promotion of the interests of the breed;
(2) maintaining breed purity;
(3) ensuring adequate pedigree records are kept;
(4) deciding on breeding objectives that can be applied to the breed as a whole;
(5) ensuring through inspection that animals registered with the society meet the 'standards' of the breed;
(6) maintaining a flock book;
(7) promoting the keeping of performance records (in some breeds).

The relative importance given these facets has varied tremendously over the years and between breeds. Traditionally emphasis has been given to 'purity' (pedigree) and breed type in decisions as to whether sheep are recognised as registered members of the breed. While this may have helped maintain the position of registered breeders within the breed hierarchy it has probably done little to aid the genetic improvement of the sheep population.

Traditionally the breeding objectives for the breed have usually been expressed in terms of the description of an ideal animal. While many of the characteristics mentioned were related to performance, production factors of high economic importance received relatively little emphasis in these objectives.

While interest is increasing in Sheeplan, only a proportion of ram breeding flocks registered with their Breed Society, with the notable exception of the Coopworth breed, are enrolled in this national recording service, or base their selection and registration on objective recording systems. There has also been, until recently, a lack of widespread demand for stock of superior performance in economically important characteristics. This situation arose partly from the absence of on-farm demonstrations of the biological and financial superiority of sheep bred by using objective selection criteria over those bred by traditional methods.

There has also been a lack of clear detailed blueprints for stud breeders to follow covering all aspects of genetic improvement. Strong social and economic forces within the industry tend to maintain the *status quo* in matters pertaining to livestock improvement. With the costs of major inputs to most breeding programmes (stock, land, labour) increasing rapidly, and new techniques being developed, there is an urgent need for guidelines to be developed by sheep societies for breeders to follow in aiming for improvement. The guidelines need to take cognizance of relevant information from several fields and adapt this information and experience to sheep improvement in New Zealand.

Breed representatives also need to use these blueprints to create an awareness of the need for change in all interrelated sections of the sheep industry. This is best done by demonstrating the economic superiority of sheep bred by objective criteria over those bred by traditional subjective methods. One of the key feature roles of the breed societies is to develop all the skills and resources needed to increase the rate of genetic improvement in ram breeding flocks of that breed, and to inform producers of the benefits of effective breeding programmes.

The diversity of selection objectives in different breeds such as the Merino, Drysdale, Coopworth and Romney, or between breeding ewe breeds and the meat breeds, and the range in productive traits likely to give the highest economic return in different farming systems and environments, justify continuation of separate societies representing each breed or breeding objective in New Zealand. There are nevertheless, several common fields of interest for all sheep breeders, including the evaluation of marketing of sheep products, research, extension, and recording services. These interests could be co-ordinated by a National Sheep Improvement Council, with assistance from the Federation of Sheep Group Breeders and Breed Societies.

The sheep industry also needs to encourage flexibility in breeding objectives so that changing production or marketing specifications and demands from commercial producers can be met. Breed Societies should open registration to, or provide separate registers for, suitable performance recorded animals to assist with this development. A

194

diversity of selection emphasis within breeds may need to be encouraged to ensure the future financial viability of ram breeders following the guidance of breed societies.

In summary, some of the key roles of breed societies in future would appear to be:

(1) To promote more direct involvement in objective recording in ram breeding flocks as an aid to selective breeding and increased rates of genetic improvement, and to encourage a greater demand for stock with superior performance in economically important characteristics.

(2) To base registration on productive performance of economically important heritable traits as well as keeping appropriate pedigree records.

(3) To encourage the collection of relevant technical information and expertise, and communicate this information to breeders and related groups.

(4) To promote the development of more effective breeding structures and the effective utilisation of animals of superior breeding value.

(5) To actively participate in the development of systems of genetic improvement that are commercially viable.

(6) To promote the adoption and co-ordination of new technology which can increase the efficiency of breeding programmes, or the more effective use of rams by commercial farmers.

With the rapid development of new technology in agriculture, and several other fields, it is obviously necessary for farmers, scientists, advisers, agricultural journalists, marketing personnel and others to integrate their experience and skills more effectively. The breed societies could fulfill an essential role in this respect under the leadership of a National Sheep Improvement Council.

REFERENCES

1 Allison, A. J.; Kelly, R. W.; Lewis, J. S.; Binnie, D. B. 1975: Preliminary studies on the efficiency of mating ewe hoggets. *Proc. NZ Soc. Anim. Prod. 35:* 83.

2 Al-Murrani, W. K. 1974: The limits to artificial selection. *Anim. Breed. Abstr. 42:* 587.

3 Bichard, M. 1971: Dissemination of genetic improvement through a livestock industry. *Anim. Prod. 13:* 401.

4 Bindon, B. M.; Piper, L. R. 1976: Assessment of new and traditional techniques of selection for reproduction rate. In *Sheep Breeding*, (Ed's G. J. Tomes, D. E. Robertson and R. J. Lightfoot). Western Australian Institute of Technology p 154.

5 Carter, A. H. 1976: Exploitation of exotic genotypes. In *Sheep Breeding*, (Ed's G. J. Tomes, D. E. Robertson and R. J. Lightfoot). Western Australian Institute of Technology, p 117.

6 Ch'ang, T. S.; Raeside, J. I. 1957: A study on the breeding season of Romney ewe lambs. *Proc. NZ Soc. Anim. Prod, 17:* 80.

7 Ch'ang, T. S.; Rae, A. L. 1970: The genetic basis of growth, reproduction and maternal environment in Romney ewes. I. Genetic variation in hogget characters and fertility of the ewe. *Aust. J. Agric. Res. 21:* 115.

8 Clarke, J. N. 1972: Current levels of performance in the Ruakura fertility flock of Romney sheep. *Proc. NZ Soc. Anim. Prod. 32:* 99.

9 Clarke, J. N. 1975: New techniques of reproduction and their relevance to breeding groups. *Sheepfmg A:* 101.

10 Clarke, J. N.; Dobbie, J. L. 1976: Selection for twinning in sheep. *Proc. Ruakura Fmrs' Conf.:* 100.

11 Dickerson, G.; Laster, D. B. 1975: Breed, heterosis and environmental influences on growth and puberty in ewe lambs. *J. Animal Sci. 41:* 1.

12 Dunlop, A. A.; Young, S. S. Y. 1961: A comparison of genetic progress in wool production under artificial insemination and natural mating. *Proc. Conf. Artificial Breeding of Sheep in Australia.* (Ed. E. M. Roberts) Univ. NSW Kensington, p 189.

13 Dyrmundsson, O. R. 1973a Puberty and early reproductive performance in sheep I. Ewe lambs. *Anim. Breed. Abstr. 41:* 273.

14 Dyrmundsson, O. R. 1973b: Puberty and early reproductive performance in sheep II. Ram lambs. *Anim. Breed. Abstr. 41:* 419.

15 Fairnie, I. J. 1976: Organisation of artificial breeding programmes of sheep in Western Australia. In *Sheep Breeding* (Eds G. J. Tomes, D. E. Robertson and R. J. Lightfoot). Western Australian Institute of Technology, p 500.

16 Hight, G. K.; Rae, A. L. 1970: Large-scale sheep breeding: its development and possibilities. *Sheepfmg A:* 73.

17 Hight, G. K.; Lang, D. R.; Jury, K. E. 1973: Hill country sheep production V. Occurrence of oestrus and ovulation rate of Romney and Border Leicester × Romney ewe hoggets. *NZ Jl Agric. Res. 16:* 509.

18 Hight, G. K.; Dalton, D. C. 1974: The role of integrated breeding groups in the sheep and beef cattle industries. *Sheepfmg A:* 128.

19 Hight, G. K.; Gibson, A. E.; Guy, P. L. 1975: The Waihora sheep improvement programme. *Sheepfmg A:* 67.

20 Jackson, N.; Turner, H. N. 1972: Optimal structure for a co-operative nucleus breeding system. *Proc. Aust. Soc. Anim. Prod. 9:* 55.

21 James, J. W. 1966: Selection from one or several populations. *Aust. J. Agric. Res. 17:* 583.

[22] James, J. W. 1976: The theory behind breeding schemes. In *Sheep Breeding*, (Eds G. J. Tomes, D. E. Robertson, R. J. Lightfoot) Western Australian Institute of Technology, p 145.

[23] Land, R. B. 1974: Physiological studies and genetic selection for sheep fertility. *Anim. Breed. Abstr. 42:* 15.

[24] Lopez-Fanjul, C. 1974: Selection from crossbred populations. *Anim. Breed. Abstr. 42:* 403.

[25] McConnell, G. R. 1974: Structure and growth of sheep group breeding schemes. *Sheepfmg A.:* 44.

[26] McDonald, M. F. 1974: The use of ram lambs as sires. *Sheepfmg A.:* 1.

[27] McDonald, M. F. 1975: Progress in the transfer of eggs between sheep. *Sheepfmg A.:* 146.

[28] Nicoll, G. B. 1976: The place of permanent large-scale breeding schemes in livestock improvement. *NZ Agric. Sci. 10:* 49.

[29] Parker, A. G. H. 1976: Advantages of Group breeding schemes. In *Sheep-Breeding* (Eds G. J. Tomes, D. E. Robertson and R. J. Lightfoot) Western Australian Institute of Technology, p 214.

[30] Rae, A. L. 1976: The development of co-operative breeding schemes in New Zealand. In *Sheep Breeding* (Eds G. J. Tomes, D. E. Robertson and R. J. Lightfoot), Western Australian Institute of Technology, p 154.

[31] Rae, A. L. 1977: Group breeding schemes in sheep improvement in New Zealand. *Proc. NZ Soc. Anim. Prod. 37:* 206.

[32] Rae, A. L.; Wickham, G. A. 1970: Crossbreeding and its part in the utilisation of existing and introduced breeds. *Sheepfmg A.:* 87.

[33] Turner, N. H. 1969: Genetic improvement of reproduction rate in sheep. *Anim. Breed. Abstr. 37:* 545.

[34] Wallace, L. R. 1964: The effect of selection for fertility on lamb and wool production. *Proc. Ruakura Fmrs' Conf.:* 1.

8
Performance Recording

D. C. Dalton

Ruakura Animal Research Station, Hamilton

Consultants: Clare F. Callow, V. R. Clark, and K. J. Dunlop

SUMMARY

The aim of performance recording is to document information which can then be used in making selection decisions. Details are discussed on effective identification of each animal, on the collection of mating records, lambing records, weight recording, fleece recording and the use of selection lists. Breed Society flock books are discussed in the light of future developments in recording. The main theme of the discussion is that unless performance records are used in making better selection decisions, then their value is limited. Sheeplan — the National Flock Recording Scheme — is used to illustrate details of performance recording.

INTRODUCTION

Performance recording in sheep has a simple, clearly defined aim — to document information which can be used to make decisions.[5] These decisions are selection decisions that are concerned with the question of which animals to keep to breed future generations, and which animals to cull. It is on these "keep or cull" decisions that the whole of genetic improvement is based.

The real challenge to a performance recording scheme is to maintain an effective data collection and processing system. This starts with information collected in the paddock through to the selection decisions made by the breeder in his office and subsequently in his sheep yards.

The two aspects of collecting and processing data are really separate. The data actually collected are past performance or history on the animal, and processing is essential to turn these data into a form in which they can be used for future predictions of the animal's genetic worth.

Performance recording has often been considered as a separate aspect to pedigree recording but, to be effective, it is essential that there is close association between the two for both administrative and genetic reasons. The recording of pedigree is of course a basic component of any performance recording scheme where the central core is the individual records of each animal.

The success of performance recording in any flock depends on

199

whether the trait to be recorded can be measured, whether it is of economic importance and whether it is heritable and will respond to selection.

In New Zealand, *Sheeplan* (the National Flock Recording Scheme) is the basis of national sheep improvement.[1] It enables breeders to identify, measure, rank and select individual sheep within a flock on four productive measurements. These are:

(1) number of lambs born or reared
(2) weaning weight
(3) autumn, winter and spring hogget weight
(4) hogget fleece weight.

SHEEP IDENTIFICATION

The basic feature of an identification system is that within the flock, each animal should be given a unique identity. This is given to the animal at birth and remains with it throughout its life.

A unique identification can be made up from two components. These are:

(1) an individual number — its tag number
(2) the year born — its tag year.

Thus, a complete identification is tag number + tag year. An example would be animal 1234, born in 1977 which is identified as 1234/77. It is only necessary to use the last two digits of the year born.

In computerised systems (such as Sheeplan), it is best to avoid using letters in the identification system, as numbers are much easier to handle in computer programming.

There is a wide range of tags and tagging systems available, depending on the requirements of each recording system. The tag may be temporary or permanent, be for close or distant reading and denote either group or individual identity.

Examples of *temporary identifiers are*:
stick-on labels or tie-on ribbons
raddle
painted numbers
neck-tags

Examples of *permanent identifiers* are:
coloured wire twisters in ear
numbered metal eartags
large plastic numbered eartags
ear tattoos
ear notches
numbers burnt on horns.

The most commonly used system is to insert a numbered brass eartag in the leading edge of the left ear at birth. This is a reliable permanent

Photo 8.1: A ewe modelling four different forms of identification, small plastic, large plastic and brass eartags, together with a neck tag.

identification which interferes least with shearing and can be easily read.[3]

Large plastic coloured and numbered tags are often inserted later at the hogget or two-tooth stage when the ear is large enough to carry the tag. These tags can be read easily from a distance, and using a range of colours can be used to denote age, sire group etc.

The main aim of the breeder is to tag for ease and accuracy of data collection. Lost tags mean lost identity and hence lost data. The insertion position of the tag in the ear is important and some breeders consider the extra work of a double tagging system justified. It is important to replace lost tags with the correct original number to avoid confusion through cross-reference from the new to the old identity.

The positioning of tags on the animal (e.g. eartags) is best dictated by the recording system. For example, the layout of the weighing race can dictate how the sheep come up to the scales and this may decide into which ear the tag is inserted. Having to twist tags and hence twist ears to read them prior to weighing adds unnecessary stress to both sheep and men. Similarly, the positioning of tags in ears to avoid damage to shearing gear is important. Tags are never popular with shearers but at least consistency in the insertion position in the ear is appreciated by most shearers.

MATING RECORDS

It is generally accepted that the sheep production year starts at the joining of the ram with the ewes. The term "joining" is best used as it then allows the term "mating" to be reserved for the actual act of mating. Thus ewes may be joined but not necessarily mated by the ram.

The purpose of mating records is to record the final outcome of a successful mating, that is, the sire of the lambs which are eventually born to the ewe.

During the breeding season, the ewe may be joined with a ram (or different rams) over a number of oestrous cycles (usually 3 or 4), and accurate records are essential to record in which cycle (i.e. to which ram) she actually conceived. In some cases, in the final cycles of the breeding season, "tail-up" rams are used to cover any ewes not pregnant and still cycling. Sometimes rams of different breeds are used to make accurate sire recognition easier.

Some breeders change rams during the breeding season, regardless of the oestrous cycle. They record the day of ram change and then determine the sire by subtracting the length of an average pregnancy (147 days) from the date of birth. This practice is subject to considerable error because of the normal variation around the mean pregnancy length.

The use of "Sire-sine" mating harnesses with different coloured crayons which are changed at each oestrous cycle (17 days) is an important aid to determine mating and subsequent pregnancy.

If oestrous and mating information is needed without pregnancy, then vasectomised (teaser) rams can be used when fitted with mating harnesses. For example teasers can be used after the final entire mating cycle to record the cycling non-pregnant ewes which can then be culled.

The main task at mating is to allocate the ewes in the whole flock to each ram. As ewe identifications are read while animals go through a crush-pen or race, then their numbers will be recorded in random order.

It is advisable to use some form of check list from this stage to ensure that animals are not missed or duplicate numbers recorded. Ewes are then allocated to rams on a designed basis — either at random within age groups (where subsequent sire analyses are needed) or on some selected basis.

The most efficient system is to have an accurate list of all animals in the flock, in numerical order within ages prior to mating. This list can then be used for the allocation of ewes to each ram prior to mating and as a check-list in the yards at mating.

As each ewe comes through the race, its identification is checked off the list and it is then given a suitable group identity (either temporary or permanent). It can be drafted into its sire group there and then, or at some later stage.

From this stage (entire joining), a record of each mating cycle can be kept. For the case where one sheet is used as an overall check list, a suggested layout is shown in Table 8.1.

TABLE 8.1: Suggested Mating Sheet and Check List

Ewe	Sire	Cycle				Notes
		1	2	3	4	
123/74		√	√			
221/74		√				
21/75		√	√	√	√	Barren

Ewes are in numerical order within age-groups and a mating at each cycle is recorded by a tick

Another system is to have a separate sheet for each ram on which the same basic cycling data are recorded.

One way to reduce the amount of paper in the sheep yards is to have all the rams and their mating groups on the one sheet of paper as shown in Table 8.2. This is prepared in the farm office prior to mating.

The following recording format can be used:
(1) tick — to show that each ewe is checked as present at joining;
(2) pencil line — through the ewe identification when mated;
(3) coloured line — through the ewe identification when remated;
(4) circle — around the ewe identification to show the ewe was not mated.

The use of different coloured pens when field recording can improve the speed and accuracy of recording.

TABLE 8.2: Suggested Mating Sheet and Check List

Sires					
1/74	4/74	9/74	17/74	21/74	26/74
2-tooths					
2/74	5/74	3/74	10/74	8/74	6/74
15/74	11/74	18/74	11/74	20/74	19/74
27/74	25/74	30/74	13/74	29/74	22/74
4-tooths					
		(as above)			
6-tooths					
		etc.			

Ewes are in numerical order within age-group and sire group

In Sheeplan, the field notebook can be used to record mating information as there are three lines on which to write the rams used within the Sire Identification Column.

Weather conditions usually require the protection of mating records from the weather. A waterproof folder such as a plastic clip-board and fold-over cover is ideal.[3]

LAMBING RECORDS

Lambing is the most important time in sheep recording as this is when actual pedigreeing is completed in most flocks. In a few "easy-care" or unshepherded flocks, pedigreeing of the lamb (that is recording its identity and its mother's identity) may be done at docking time (4-6 weeks old).

Accurate dam identity is critical to all future records and their processing. The chance of errors increases considerably with multiple births and increased concentration of ewes and lambs in the lambing paddock. The shepherd can often find it difficult to determine which lambs belong to which ewes. In flocks with high multiple birth rates, some shepherds put a raddle mark on sets of twins and triplets as soon as possible after birth, before they start their tagging-round some time later in the day.

At lambing, a great deal of information other than lamb and dam identification can be recorded. This is the time when sex of lamb, birth weight, litter size, lamb mortality, lambing difficulty and maternal behaviour can also be recorded.

Much of this information is most easily recorded in coded form and examples in Table 1 are those used by Sheeplan.

TABLE 8.3: Codes Used in Sheeplan to Record Field Data

EWE FATE CODE	EWE CODED REMARKS
1 = Dead	A = Assisted
4 = Cull	B = Aborted
5 = Sold	C = Foster Mother
6 = Missing	D = Screened
7 = Barren - cull	
8 = Barren - retain	

LAMB FATE CODE	BIRTH-REARING RANK
1 = Dead - unspecified	1 = Single
2 = Dead - lamb's fault	2 = Twin
3 = Dead - ewe's fault	3 = Triplet
4 = Cull	4 = Quadruplet, etc.

LAMB CODED REMARKS	SEX
F = Fostered lamb - ewe's fault	R = Ram
G = Fostered lamb - other reasons	E = Ewe
H = Dead at birth	
I = Died subsequently	

Lambing information is usually recorded in stages. The recommended minimum is a two-stage system because it reduces the risk of errors caused by transcribing data from one source to another. This consists of the following:

(1) A field note book. This is a very important item and should be designed for easy use in the paddock in a range of weather conditions. It can be prepared before lambing by entering the dam's identification in numerical order within age-groups in the front half of the book. It should then have provision for a lamb list in numerical order at the back of the book. At lambing, in the paddock, the lamb's sex and dam identification is added to this lamb list. Thus the completion of front and back of the book forms a cross-reference system and should be kept up-to-date during lambing.

(2) A permanent record which can also serve as an input for computerised systems.

An example of the data which can be recorded at lambing is that used in Sheeplan. This is as follows:

Sire	Identification	
Dam	Identification	Date lambed
	Fate code	Coded remarks
		Breeder's remarks

Lamb	Identification	Rearing rank
	Fate code	Coded remarks
	Sex	Breeder's remarks
	Birth rank	Sire summary coded remarks
Remarks	Carried forward	
	Not carried forward	

Wet weather at lambing time causes most trouble to recording performance data, and breeders have developed methods to alleviate some of these problems. Examples are, writing in field record books inside a robust plastic bag and carrying a towel around the neck to dry hands before writing. In continuous rain some shepherds write on white painted hardboard with a pencil. Some have used tape recorders and transcribed the data later. Other suggestions have been documented.[3]

WEIGHT RECORDING

Weight records are an important part of a performance recording scheme because live weight is an important trait as such, and it is also important because of its indirect effect on other traits such as reproduction. Apart from their importance as a genetic trait, weight records are very valuable in assisting with management decisions.

There are two major sources of error in recording live weight of sheep. The first is "gut-fill" which is a common phenomenon of all ruminants. The other is the wetness of the fleece. The ideal is to weigh sheep which have been starved overnight and have dry fleeces (especially bellies). However, this is not always practical so attempts are usually made to weigh sheep full, and as quickly as possible at each weighing. The practical way out is often to accept gut-fill and wet sheep, but make sure that all animals are at the same stage of fullness and wetness when weighing.

Birth weights, if recorded, can be obtained by using a pocket spring balance hooked around the lamb's hind leg. Later and heavier weights (e.g. weaning, hogget and ewe weights) are best recorded on the various types of scales which are available.

FLEECE RECORDING

Fleece recording is concerned with two main aspects of wool production. The first is fleece weight (i.e. greasy fleece weight) and the second is fleece quality which is made up of a number of components.

Fleece weight is a clearly defined trait and breeders are encouraged to give major emphasis in their selection programmes to this trait.[4]

Fleece quality on the other hand, covers a wide range of both objective and subjective criteria, shown in Table 8.4.

TABLE 8.4: Objective and Subjective Aspects of Fleece Quality

Objective	Subjective
Staple length	Quality number
Crimps per cm	Tenderness or break
Staple strength	Medullation
Fibre diameter (micron)	Colour (discolouration)
Clean scoured yield	Bulk
Medullation (Benzol test)	Cotting
Colour (degree of whiteness)	Character
Bulk	Style
	Handle
	Lustre
	Resilence
	Pigmentation

In practice, the fleece information recorded varies greatly, depending on whether it is done for marketing reasons or for genetic reasons. For example the entire clip may be classed into broad divisions on quality number (coarse, medium, fine), or sheep may be drafted into long and short staple-length prior to shearing. These are marketing practices. For genetic reasons, fleece weight, fibre diameter and yield may be recorded.

Recording fleece information usually poses few practical problems as it is usually carried out under cover (woolshed or covered yards) when sheep are dry. Fleece weights are recorded on a wide range of suitable scales after the fleece has been picked up from the shearing board, prior to throwing on the wool table. Belly wool may or may not be included in this weight. Some wool tables have scales incorporated.

The main practical problem is to remove the fleece from the shearing board, with the correct identity of the sheep from which it came clearly recorded in some form. Various techniques are used, from stick-on labels to separate record sheets for each shearer. Check lists are often used by breeders or the data can be transcribed directly on to the more permanent record or computer input-sheet.

If more detailed wool data are required, then a sample is taken from the midside of the thrown fleece on the wool table (about 50 g) for laboratory analyses.

SELECTION LISTS

The main "keep or cull" decisions in a flock recording scheme are made on the replacement two-tooth rams and ewes before they enter the flock. These decisions are made on what are called the two-tooth selection lists. On these lists all the available information on the individual animals and their background should be assembled for ease of decision making.

Selection lists for the two-tooth rams can be used to select the flock sires, as well as acting as a sale catalogue for buyers' inspection before purchasing their rams.

Selection lists can contain large amounts of information, and the main practical problem is for the breeder, or more especially the ram buyer, to determine which data he should use to make his selection. Information is available on the sire of the two-tooth, the dam and the two-tooth itself.

Sheeplan uses the concept of *breeding values* (See Chapter 4) to help in decision making.[2] These breeding values are expressed as deviations from the mean and the units are either in units of the trait (e.g. number of lambs or kg of weight), or in cents where many traits are combined and scaled according to their relative economic values.

Selection lists can be produced in animal identification order or in some other order based on individual merit to make selection of the top animals easier. Sheeplan for example, produces two-tooth selection lists in identification order (mandatory) and then in index order for lamb production as an optional extra.

The other important selection list in a performance recording scheme is a ewe summary. A ewe summary ideally should show the updated performance of each individual ewe in the flock, so that "keep or cull" decisions can be made on the entire ewe flock. In Sheeplan the ewe summary gives information on lamb production, fertility, mothering ability, wool, progeny and pedigree information. The concept of breeding values is again used and it is updated each year. Overall flock and within-age-group averages allow comparisons of individuals with others in the age group. A ewe summary can also be produced in merit order for each ewe using lamb production as the main trait.

BREED SOCIETY FLOCK BOOKS

The flock books of Breed Societies and Breed Associations have had a similar format almost since the time of their first volumes. The basic information provided includes the following:

Names of Officers and Council
List of members' names and addresses (alphabetical order with flock number)

List of single-entry rams against owners (in alphabetical order)
Rules of the Society
Flock information
List of registered stud prefixes
List of official judges and inspectors

The information on each individual flock is generally in a standardised format and covers the following:

Name and address
Telephone number
Flock number
Breeder's stud prefix
Details of flock history
Details of animals registered — from the previous year and in the current year
List of single entry rams used in the flock in recent years.

In New Zealand some breed societies still have an open flock book where commercial ewes which meet the specifications of the society can be registered. The specifications can be performance data or a visual inspection, or both. Some societies retain the power to de-register sheep which do not meet their specifications. Other societies have a permanently closed flock book where ewes can only be registered if bred from registered parents, regardless of performance.

Interest has been expressed recently by some breed societies in incorporating performance data in their flock books. At the present stage, discussions of broad outlines and problems are taking place. The important points concerned in merging pedigree and performance data include the following:

(1) The design of a system to merge the performance data from the centralised scheme (Sheeplan) with each separate breed society office. Consideration should perhaps be given to centralisation of all breed society's data to make such overall data combination easier.

(2) Decisions have to be made as to what information is presented in the flock book for members who are performance recording. The greater danger is in presenting absolute mean values of performance rather than say deviations from the mean within each flock. If mean values are produced it confounds the farm effects (environment) with the genetic potential of each flock.

(3) For breed societies who make performance recording mandatory the problems are somewhat different. As Sheeplan for example, publishes its computer output of users' names and addresses with their chosen options, it may be easier to add flock data to this list so that a flock book of different format is produced.

FUTURE DEVELOPMENTS IN RECORDING

The future of sheep production in New Zealand is clearly going to require greater emphasis on genetic improvement. This means more recording for genetic reasons. However, economic pressures will also bring about an increased interest in simplified recording schemes for sheep producers who are not classed as breeders (either stud or commercial). This latter trend will certainly take place if there is a move toward greater intensification of sheep enterprises to exploit sheep of high potential as opposed to systems based on the extensive easy-care concept.

Increased recording will require improvements in practical techniques which reduce labour and increase accuracy of data collection.[3]

It will also require improvements in computing systems. Sheep data, being seasonal, require sporadic data punching processes and hence clarity and accuracy of field data greatly assist a data punch pool working under pressure. Developments are required to record data in a form which can reduce the punching load, for example recording weights direct on to tape.

The computerisation of records is an accepted essential for the future. Breeders in future cannot rely on old systems such as individual record cards and ledgers, as these preclude any possibility of quick and accurate manipulation of data to convert it into a form which can be used in decision making. Unless performance recording results in better breeding decisions, then the expense and effort involved is not justified.

REFERENCES

[1] Dalton, D.C.; Callow C. F. 1976. Why Sheeplan is necessary. *Proc. Ruakura Frmrs' Conf.* 13-17.

[2] Clarke, J. N.; Rae, A. L. 1977. Technical aspects of the present national sheep recording scheme (Sheeplan). *Proc. N.Z. Soc. Anim. Prod. 37:* 183

[3] FLBG. 1975. *Proceedings of the 2nd Conference. New Zealand. Federation of Livestock Breeding Groups, Wellington.* 152pp

[4] NZSAP. 1974. *Guidelines for wool production in New Zealand.* Occasional Publication No. 3 New Zealand Society of Animal Production, 40pp

[5] Owen, J. B. 1971. *Performance recording in sheep.* Technical Communication No. 20. Commonwealth Bureau of Animal Breeding and Genetics. Edinburgh. 132pp.

9
Principles of Reproduction

J. F. Smith

Ruakura Animal Research Station, Hamilton
Consultants: *R. W. Kelly, K. R. Lapwood* and *H. R. Tervit*

SUMMARY

An appreciation of the principles of reproduction is necessary for an understanding of the mechanisms of the factors that can affect the efficiency of the reproductive process. This chapter outlines the anatomical, physiological and endocrinological factors that control the reproductive process of the ram and the ewe and discusses them in relation to the various factors that affect the efficiency of reproduction in New Zealand sheep flocks.

STRUCTURE OF THE RAM'S SEXUAL ORGANS

General

The reproductive organs of the ram are shown in Fig. 9.1 while the detailed arrangement of the testicular and epididymal tubes is shown in Fig. 9.2.

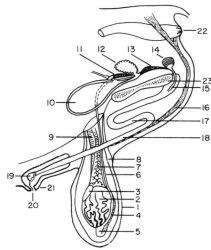

Fig. 9.1: General anatomy of reproductive organs of the ram

1. Scrotum **2.** Testis **3.** Head of epididymis **4.** Body of epididymis **5.** Tail of epididymis **6.** Vas deferens **7.** Pampiniform plexus **8.** Tunica dartos muscle **9.** Cremastor muscle **10.** Bladder **11.** Ampulla **12.** Seminal vesicle **13.** Prostrate gland **14.** Bulbo-urethral gland **15.** Urethra **16.** Retractor penis muscle **17.** Sigmoid flexure **18..** Penis **19.** Glans penis **20.** Urethral process **21.** Prepuce **22.** Anus **23.** Pubis bone.

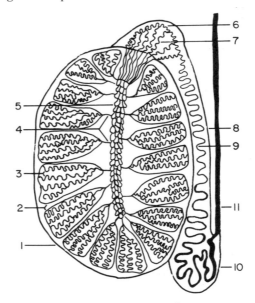

Fig. 9.2: Anatomy of testis

1. Tunica albuginea **2.** Lobules of parenchyma **3.** Seminiferous tubules **4.** Straight tubules **5.** Rete testis **6.** Head of epididymis **7.** Efferent tubes **8.** Body of epididymis **9.** Epididymal duct **10.** Tail of epididymis **11.** Vas deferens.

The scrotum, in which the paired testes are contained, supports and protects the testes, and has an important role in regulating testicular temperature. The testes produce spermatozoa and the male sex hormones. They are relatively large in the ram, weighing up to 300 gm each. Within the scrotum the testis is covered by a tough fibrous membrane *(tunica albuginea)* which contains the testicular arteries and veins. Inside this membrane is the testicular parenchyma consisting of cone shaped lobules, each containing several thin, long, tortuous seminiferous tubules. These unite into straight tubules in the centre of the testis and connect to a network of ducts *(rete testis)* joining via the efferent tubules to the head of the epididymis. Between the seminiferous tubules are blood vessels, nerves and interstitial tissue comprised mainly of Leydig cells.

The elongated epididymis is closely adherent to the testis. It comprises a flattened U-shaped head on the upper region of the testis leading to a long thin body which expands into a larger rounded sac (the tail) on the bottom of the testis. The head of the epididymis contains several convoluted tubes connected to the rete testis. These tubes unite to form a single epididymal duct of up to 60 m length and very convoluted within the body and tail of the epididymis. The epididymis is involved with the transport, maturation and storage of spermatozoa. The spermatozoa are moved from the tail of the

212

epididymis to the penis through the *vas deferens*. The last 3 - 4 cm of the two deferent ducts are thick and form the *ampullae*.

A group of accessory sex glands are situated in the pelvic cavity in close association with the ampullae. These glands produce fluid secretions (seminal plasma) in which the spermatozoa are transported through the urethra. Most of the seminal plasma is produced from the pair of vesicular glands (seminal vesicles).

The urethra, which serves a dual function of transport of semen as well as expulsion of urine, runs through the penis or male copulatory organ. The penis is covered with a white and dense fibrous membrane and this is filled with cavernous tissue connected with special blood vessels. Upon sexual excitement the spaces in the cavernous tissue become filled with blood and the penis becomes rigid. During copulation the penis becomes lengthened through the straightening out of the sigmoid flexure. The penis has many nerve endings and its most sensitive part is the head or *glans penis*. There is a 3 - 4 cm twisted urethral process (or filiform appendage) attached to the glans penis and during ejaculation this rotates rapidly and sprays semen about the anterior vagina of the ewe. Except during mating, the end of the penis is protected by a sheath of skin, the prepuce.

Development

Gonadal differentiation of the testis is apparent histologically about the 35th day of foetal life and testicular descent into the scrotal pouch is completed by the 80th day. In the sexually developing ram, testicular growth follows a sigmoid-shaped curve, with a slow rate in the first 2 - 3 months after birth, a rapid phase between 2 - 5 months when sperm production is becoming established, then a subsequent slowing of growth.

The phase of rapid growth is more closely correlated to body weight than to age[9] and corresponds to the period when body weight increases from 22 kg to 28 kg[36] although there are marked breed differences. The development of the reproductive tract and accessory organs is highly correlated with testicular growth, largely as a result of the increased secretion of sex hormones which occurs before and during puberty.

Structure of Spermatozoa

A spermatozoon consists of two parts, the head and tail (Fig. 9.3). The head is a flattened ovoid part containing the nucleus and the anterior portion is capped by an envelope or acrosome.

The tail is long and flagellum-like and is concerned with movement of the sperm. The tail consists of a number of contractile fibrils (axial core) and comprises three regions: (i) the mid-piece which is thicker

213

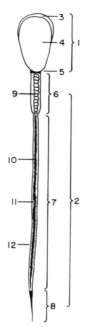

Fig. 9.3: Structure of Spermatozoa

1. Head **2.** Tail **3.** Acrosome **4.** Nucleus **5.** Neck **6.** Mid piece **7.** Main piece **8.** End piece **9.** Mitochondrial sheath **10.** Axial core **11.** Fibrous sheath **12.** Lipoprotein membrane.

and is surrounded by a mitochrondrial sheath, (ii) a main piece covered by a fibrous sheath, and (iii) an end piece which is short and not enclosed by a sheath.

The surface of the sperm cell is covered by a lipoprotein membrane.

PHYSIOLOGY OF SPERM PRODUCTION

Spermatogenesis

Spermatozoa are formed in the seminiferous tubules of the testes (Fig. 9.3). In immature rams, the inner walls of these tubules consist of the sperm-producing epithelium which contains numerous spermatogonia and undifferentiated supporting cells; the latter proliferate to form the Sertoli cells of the adult testes. Fig. 9.4 illustrates the spermatogenic cycle. Spermatogonia in the testicular tubules develop first into A1-spermatogonia, which constitute the stem cells of the spermatogenic series. They have two roles, to maintain the supply of resting spermatogonia and to initiate the series of divisions which result in production of mature sperm. During spermatogenesis the chromosome number is halved at the division of the spermatocytes. There are no further cell divisions once the spermatids are formed.

214

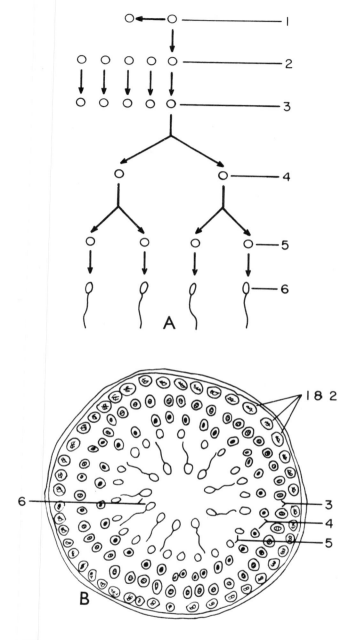

Fig. 9.4: (A) Spermatogenesis (schematic outline)
(B) Seminiferous tubule (cross section)

1. A-1 spermatogonia **2.** Spermatogonia **3.** Primary spermatocytes (2n) **4.** Secondary spermatocytes **5.** Spermatids (n) **6.** Spermatozoa.

The spermatids are transformed by a series of changes into spermatozoa.[25]

Generally a stage of the cycle occupies an entire cross-section of the seminiferous tubule. Stem cell division takes place regularly and does not wait until previous generations of spermatozoa have been released into the lumen of the tubula.

The total duration of spermatogenesis is about 49 days while the period between successive appearances of the same cycle at a particular point in a tubule is ten days.

Time of Onset of Spermatogenesis

The time of onset of spermatogenesis, when primary spermatocytes first appear in the tubules is more closely associated with body weight than with age. Typically A1-spermatogonia appear at about a body weight of 21 kg, and all spermatogenic cell types are present about 27 kg, although there are breed differences in the exact body weight at which each stage commences.[9] Maximal efficiency of spermatogenesis does not occur until several months after the first appearance of spermatozoa; partly due to variation in time of onset of spermatogenesis in different seminiferous tubules and partly because an initial high rate of degeneration of intermediate spermatogonia decreases with age.

Sperm Production and Transport

Estimates of the daily sperm output of rams have been highly variable ranging from 5.5 to 13.9×10^9 sperm/day. This variation has been attributed to technical variations in methods of estimation, as well as to breed, age, season and nutrition and their interactions.

After the sperm have been shed into the lumen of the seminiferous tubules they are transported, by the movement of fluid secreted by these tubules, through the straight tubules and rete testis, into the head of the epididymis.[35] The newly formed spermatozoa in the seminiferous tubules are immature, immotile and very sensitive to unfavourable climatic and nutritional factors. They remain in the epididymis for a period of 11 to 14 days. This period varies slightly according to frequency of ejaculation. While they are in the epididymis the spermatozoa mature and become more motile and resistant to heat and cold. Most importantly they acquire the ability to fertilise.

The epididymal tail acts as a reservoir for sperm. There is a relatively constant supply of sperm from the testis, with a variable overflow through the vas deferens into the urethra and voiding in the urine influenced by sexual activity.

Ejaculation

The process of ejaculation involves a series of coordinated contractions which begin in the epididymis, and pass along the vas deferens. Simultaneous contractions of the accessory glands mixes spermatozoa from the ampullae with accessory gland fluid in the urethra. Contractions of the urethral muscle expel the semen via the urethral process. Ejaculation in the ram takes place in less than one second.

Temperature Regulation

The testes produce sperm most efficiently when their temperature is 4 to 7°C lower than the body temperature.[24] Infertility can result from the failure of the ram to maintain this optimum temperature. Several mechanisms are involved:[35] (i) The scrotal skin contains numerous large sweat glands and evaporation of their secretions cools the scrotum; (ii) The scrotal skin is thin and lacks subcutaneous fat which allows the superficial blood vessels to be closer to the surface, and thus blood leaving the scrotal surface is cooled considerably; (iii) Two muscles (the tunica dartos and external cremaster) contract in cold conditions and force the testes closer to the abdominal wall for warmth. Under hot conditions the muscles relax and by allowing the scrotum to become pendulous provide greater surface area for loss of heat through sweating and evaporation; (iv) Finally there is a heat exchange mechanism operating between the arteries entering and veins leaving the testes. In the spermatic cord the spermatic artery is long and coiled and lies in close proximity to the very extensive venous network through which blood returns to the body. In this region the arterial blood is cooled while the venous blood is warmed by several degrees.

HORMONES OF RAM REPRODUCTION

The pituitary gonadotrophins, follicle stimulating hormone (FSH) and luteinizing hormone (LH), are required for the multiplication of the spermatogonia and supporting cells of the impuberal testes.[7] Male sex hormones (testosterone) may also be involved at this stage and in the formation of A1-spermatogonia. Further divisions of the spermatogonia and the formation of the spermatocytes do not require either gonadotrophic nor gonadal hormones. However testosterone appears necessary for the reduction division to spermatids while FSH is required for the final stages of spermatid maturation.

The development and function of the accessory glands, growth of penis and scrotum and expression of mating behaviour (libido) are dependent on testosterone production by the Leydig cells of the testes. Secretion of testosterone commences at about the time of differentiation of the testis (i.e. day 30-35 of foetal life). Prenatal testosterone production and testicular growth may be influenced by placental gonadotrophin production. There is a fall in plasma levels of both LH and testosterone during the last three weeks of gestation and for about the first seven days post-partum, possibly due to a feedback effect of maternal oestrogen production just before parturition. Levels of FSH, LH, and testosterone increase shortly after birth. The levels of FSH are maximal at about five weeks while LH appears to peak at about 80 days and then decline while testosterone levels continue to rise. The negative feedback mechanism (i.e. increase in level of one hormone results in a decrease in the level of another) between the testis and the hypothalamic-pituitary axis develops greater sensitivity at about 80 days. Testosterone production by the Leydig cells and oestradiol production by the Sertoli cells and/or Leydig cells are needed to maintain the negative feedback on LH, whereas the negative feedback on FSH is mainly due to oestradiol and to a component of the semen itself called "inhibin".

In adult rams LH and FSH are secreted in pulsatile patterns, while peaks of testosterone secretion follow 30 - 60 min after each LH peak. There does not seem to be any circadian pattern of secretion of these reproductive hormones. However marked seasonal patterns of production are seen with LH output elevated in mid to late-summer while the seasonal peak of testosterone is found in early autumn. Changes in daily photoperiod appear to be the major regulatory factor.

COMPOSITION OF SEMEN

The characteristics of semen differ markedly amongst species with the ram producing an ejaculate with a small volume but a very high concentration of spermatozoa. The chemical composition of and techniques for collection and evaluation of semen are well documented.[19][32]

Semen consists of two major fractions, the spermatozoa and the seminal plasma, which is the mixed secretions of the epididymis and the accessory sex glands. The seminal plasma has four functions; (a) to act as a vehicle for conveying the spermatozoa during ejaculation, (b) to serve as an activating and buffering medium for the previously non-motile spermatozoa, (c) to provide the spermatozoa with nutrient secreted by the accessory glands and (d) provide some form of protection of the tract from infection (i.e. ubiquitin).

Relationships of Testis Volume and Sperm Production

Although the relationship between the amount of testicular tissue and sperm production has been known for some time, the practical

218

application of this knowledge has been limited by the failure to establish relationships with parameters that can be measured in the live animal. However, recent data[14] show that there is a high correlation between testes volume and the number of sperm in the testes and epididymis. Thus selection of rams on testes volume may prove a useful indicator of their ability to cover a large number of ewes.

THE REPRODUCTIVE ORGANS OF THE EWE

The reproductive tract of the ewe is illustrated in Fig. 9.5. The gonads or ovaries are two oval shaped structures, which are approximately 1 - 2 cm in diameter, located in the abdominal cavity. They are supported by broad ligaments to the uterus and abdominal wall. The ovary is the source both of ova and the female sex hormones (oestrogen and progesterone).

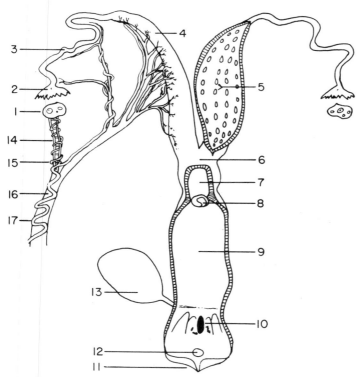

Fig. 9.5: General anatomy reproductive organs of the ewe

1. Ovary **2.** Infundibulum **3.** Fallopian tube **4.** Uterine horns **5.** Caruncles **6.** Uterine body **7.** Cervix **8.** Cervical folds **9.** Vagina **10.** Urethral orifice. **11.** Vulva **12.** Clitoris **13.** Bladder **14.** Ovarian vein **15.** Ovarian Artery **16.** Utero-ovarian vein **17.** Utero-ovarian artery.

Closely associated with and generally surrounding the ovary is the enlarged end of the fallopian tube *(infundibulum)* which at ovulation surrounds the ovary assisting the passage of the ovum into the fallopian tube.

The *fallopian tube* or oviduct is 10 - 12 cm long in the ewe and comprises the ampulla, the first third of the tube closest to the infundibulum and the remaining, somewhat smaller diameter, isthmus. The fallopian tube is lined with ciliated cells which facilitate the movement of ova towards the uterus. The isthmus of the fallopian tube joins the tapered end of the uterine horn forming the utero-tubal junction. This junction is under hormonal control and can act as a temporary barrier to the entry of the ova into the uterus.

The *uterus* is comprised of two horns about 10 cm in length (i.e. bicornate, see Fig. 9.5) which join to form a short single uterine body 1 - 2 cm in length immediately before the cervix. The walls of the uterus are comprised of two major layers, the outer muscular layer or (myometrium) and the inner glandular layer (endometrium). The inner surface of the endometrium has a large number of elevated areas or caruncles, which are the areas where placental attachment between the foetal membrane and the uterus occurs. This type of placentation is referred to as cotyledonary placentation and is characteristic of all ruminants.

The *cervix* is a cartilaginous structure approximately 5 cm in length with a very tortuous convoluted lumen. The opening of the cervix into the vagina is masked by a number of folds of tissue. The prime functions of the cervix are to act as a reservoir for spermatozoa following mating, and to protect the foetus from the external environment.

The *vagina* is some 6 cm in length and is lined with a mucosal membrane. It has an annular constriction just cranial to the opening of the urethra.[5] This constriction is more pronounced in young non-parous ewes.

The arterial supply to the tract is via three branches from the aorta: (i) the small utero-ovarian artery which supplies the ovary and fallopian tubes and tips of uterine horn; (ii) the larger mid-uterine artery which supplies the majority of the uterus, and (iii) another large vessel which supplies the body of the uterus, cervix and vagina. The venous drainage is generally via the large utero-ovarian vein which drains the uterus, ovary and fallopian tubes. The utero-ovarian artery is very convoluted and is wrapped around this major vein. This physical proximity (Fig. 9.5) has been related to certain physiological changes in the cycle and is a possible mechanism for hormonal action.

PHYSIOLOGY OF REPRODUCTION IN THE EWE

Oogenesis

Oogenesis comprises the formation, development and maturation of the female gamete. The process begins in early embryonic life (about day 35) with rapid multiplication in the ovary of the primordial germ cells and oogonia by mitotic division, which continues up to day 80 - 90. About day 52 the oogonia commence the first of the two divisions in meiosis to become oocytes. They are then surrounded by a single layer of epithelial cells and become primordial follicles. The lifetime stock of primordial follicles is established by day 100 in the sheep embryo. There are many thousands of oogonia but from about day 72 there is an onset of degeneration of the oogonia which continues through to birth.[21] Thus, unlike the ram in which the production of spermatozoa in the testis is a continuous process after puberty, the female lamb at birth contains in her ovaries all the ova available for her reproductive life.

Once the oocytes have reached the diplotene phase of the first division they enter a prolonged resting phase and further oocyte division is halted until animals approach puberty.[3 4] The epithelial cells forming the primordial follicle soon begin to multiply while retaining an orderly wall-like arrangement around the oocyte (Fig. 9.6). When this wall is some six cells thick a differentiation into an external theca and an internal granulosa takes place. The thecal cells become vascularised and continue to grow while the granulosa does not become vascularised and division is much slower. This differential growth rate is responsible for the formation of a fluid filled cavity or antrum within the granulosa which accounts for most of the volume of the mature follicle. Such a mature follicle is referred to as a Graafian follicle (Fig. 9.6 illustrates the stages of follicular development). The process of development of vesicular follicles (i.e. stage 4 or 5) from primordial follicles is constantly taking place in the ovary with the first appearing soon after birth. The number of follicles at each stage fluctuates only slightly and the population can be considered to be in a steady state. Thus the great majority of vesicular follicles that develop and all those before puberty undergo degeneration at various stages of growth. The stage of pre-ovulatory growth however is controlled by the circulatory levels of gonadotrophins.

During the maturation process meiosis resumes in the oocytes and at the time of ovulation the first division has been completed to produce a secondary oocyte with a metaphase II nucleus and the first polar body. The completion of the second division and production of the second polar body occurs at the time of fertilisation. Thus a major difference between spermatogenesis and oogenesis is that in the former, four spermatozoa are formed as a result of the divisions, while in the latter, only one mature ovum is produced. The oocyte reaches its maximum size before the antrum of the follicle develops and about this time a non-cellular layer (mainly mucopolysaccharide), the zona pellucida, forms around the oocyte and eventually acts as a protective layer for the ovum after ovulation has occurred.

Ovulation and Development of the Corpus Luteum

During the final phase of pre-ovulatory maturation a small area of the wall of the Graafian follicle and the overlying ovarian cortex becomes thin and translucent to form the "stigma". This is thought to be due to necrotic changes in the granulosa due to action of certain proteolytic enzymes. The stigma becomes raised and resembles a blister on the surface of the ovary. With the onset of ovulation this blister ruptures and the follicular fluid gently oozes out. The process is not an explosive one and the trickle of the somewhat mucous fluid which contains the now free floating ovum continues for a short time. The ovum is directed towards the fallopian tube by ciliary currents within the enveloping infundibulum. In the ewe ovulation is "spontaneous" in that it occurs about 25-30 hours after the onset of oestrus regardless of whether or not mating has occurred.

After rupture of the follicle and release of the ovum the follicle collapses and the granulosa becomes folded. It rapidly develops a vascular supply and the cells hypertrophy. This process, referred to as luteinisation, forms the corpus luteum. The corpus luteum grows rapidly during the first 6 - 8 days after ovulation and at the same time increases its production of the hormone progesterone. The life span of the corpus luteum is dependent on whether or not the ovum becomes fertilised and implants. If conception occurs the corpus luteum persists for almost five months throughout the duration of pregnancy. However if conception does not result then the corpus luteum will commence to regress about 14 days after ovulation and another ovulation will occur at the following oestrus.

Oestrus (heat) and Endocrinology of the Oestrous Cycle

Oestrus is the period of sexual receptivity in the ewe and is the time when she will accept service by the ram. The duration of oestrus, normally about 24 hours, varies from four to 72 hours depending on a number of factors such as age, and breed of ewe and degree of contact with the ram.

The onset of behavioural oestrus is associated with an increase in the levels of oestrogen in the blood which also produces a number of changes in the reproductive tract. There is an increase in the blood supply and the secretory activity of the glandular material in the uterus and vagina, both of which become engorged and turgid. An increase in

222

the secretion of mucus from the cervix occurs which is associated with changes in the consistency of the mucus throughout oestrus. At the onset of oestrus cervical mucus is clear and produced only in small amounts, but after about 12 hours secretion has become much more copious and slightly cloudy and by 24 hours has become creamy in colour and consistency.

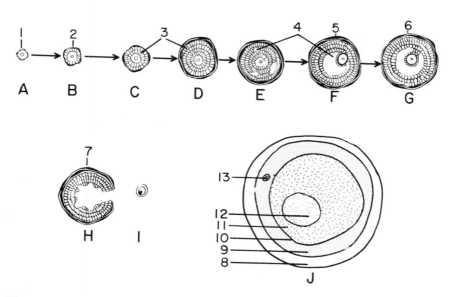

Fig. 9.6: Diagrammatic representation of follicle growth and structure of an ovum

A. Stage I **B.** Stage II **C.** Stage III **D.** Stages IV and V **E.** Stage VI **F.** Stage VII **G.** Stage VIII **H.** Stage IX **I.** and **J.** Ovum **1.** Primordial follicle **2.** Oocyte **3.** Granulosa cells **4.** Antrum **5.** Graafian follicle **6.** Theca cells **7.** Early Corpus luteum **8.** Zona pellucida **9.** Perivitelline space **10.** Vitelline membrane **11.** Vitellus (cytoplasm) **12.** Nucleus **13.** First polar body

The ewe does not have conspicuous behavioural signs of oestrus and rams detect the oestrous ewe primarily by the odour of the vaginal secretions, but also by sight. However, many ewes will actively seek out a ram and compete with other ewes for his attention. It is common for a number of ewes to form a "harem" around a ram.

The duration of the oestrous cycle or interval between the two successive oestrous periods in the ewe varies slightly among breeds and among animals within a breed but is generally about 15 to 18 days with a mean of 17 days. The cycle is characterised by a series of events that are under the control of the pituitary and ovarian hormones. The pattern of changes over the cycle for some of these hormones is illustrated in Fig. 9.7.

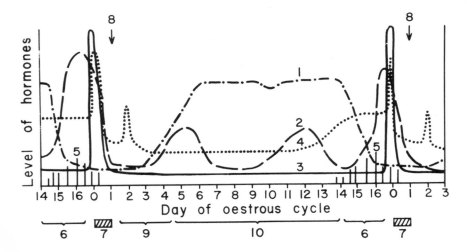

Fig. 9.7: Oestrous cycle and outline of hormonal changes

1. Progesterone **2.** Oestrogen **3.** L. H. **4.** F.S.H. **5.** Prostaglandin $F_2\alpha$ **6.** Luteal regression **7.** Oestrus **8.** Ovulation **9.** Early luteal phase **10.** Mid-luteal phase.

The most consistent part of the cycle is the luteal phase (or period of an active corpus luteum) and this largely governs the length of the cycle. This phase is characterised by the higher levels of blood progesterone (approx 3 ng/ml plasma) produced by the luteal cells. Progesterone levels are low (0.5 ng/ml plasma) at oestrus and remain below 1 ng/ml until about day 3 to day 4 of the cycle (oestrus = day 0), then they rise to reach peak values about day 7. The levels plateau at about 3 ng/ml until approximately day 14 of the cycle when they begin to fall rapidly to low levels typical of oestrus by day 16. This fall in progesterone results from the regression of the corpus luteum caused by the release of prostaglandin $F_2 \alpha$ (PGF) from the uterus. Prostaglandin appears to be synthesised and released from the uterus after the uterus has been exposed for a period to high levels of progesterone. In fact the time of the regression of the corpus luteum can be shortened by the injection of progesterone into the ewe during the first four days after oestrus.

In addition the falling levels of progesterone, while initiated by a release of prostaglandin, stimulate further release of PGF and consequently a further reduction in progesterone levels. Coincident with the fall in progesterone there is a rise in the levels of oestradiol - 17β (E2) produced by the developing follicle in the ovary. Peak levels of oestrogen (10 - 20 pg/ml) are reached about 12 hours before onset of oestrus. The oestrogen appears to be produced in response to FSH released from the pituitary gland. This release is controlled by the removal of progesterone which exerts a negative feedback effect in the pituitary. The increased levels of E_2 exert a positive feedback on the pituitary via the hypothalamus where it stimulates the production of gonadotrophin releasing hormone (LH-RH). The increase in LH-RH results in the release of a surge of gonadotrophins with sharp peaks of both FSH and LH being found about four hours after the onset of oestrus. This surge is responsible for the final maturation of the follicle and ovulation which occurs about 20 hours after the LH peak. The levels of oestrogen fluctuate throughout the oestrous cycle and three to four distinct peaks have been reported at about four day intervals. However as these occur in presence of high levels of circulating progesterone the feed back effect of oestrogen is blocked. Small episodic discharges of FSH and LH occur throughout the cycle.

The presence of an embryo in the uterus blocks the release of PGF from the uterus and thus prevents regression of the corpus luteum. The mechanism by which the embryo exerts this influence is unknown, although some immunological process may be involved. The levels of progesterone increase slowly during the next 80 days and the level of progesterone increases with the number of embryos (i.e. increased number of corpora lutea).

The number of corpora lutea formed is influenced by a range of factors. However, the hormonal changes during the last four days of the cycle (Day 13 to Day 17) seem to have the controlling influence. Depending on the amount of circulating gonadotrophins, a variable number of selected follicles mature. The selection process is not known but appears to be associated with a coordination of the stage of pre-antrum growth and the circulating levels of FSH. The stimulus of FSH and its accumulation into the follicular fluid causes a change from androgen to oestrogen synthesis by the follicle. The increased oestrogen associated with the FSH stimulates follicular growth and further oestrogen synthesis and secretion into the blood. The increasing plasma levels of oestrogen inhibit further FSH production and stimulate the development of LH receptors within the follicle as well as LH secretion by the pituitary. The increased absorption of LH by the follicle eventually leads to ovulation and luteinisation.[18] [21]

Transport of Spermatozoa and Fertilisation

During mating spermatozoa are deposited into the anterior part of the vagina. A small proportion of the spermatozoa ejaculated by the ram penetrate the cervical mucus and pass into the cervix. The cervix acts as a reservoir for the spermatozoa as the cervical environment, in comparison with other parts of the tract, enhances survival of the spermatozoa (up to 78 hours). Once a cervical reservoir has become established, spermatozoa are constantly released into the uterus and the fallopian tubes, although the utero-tubal junction may also act as a partial barrier. The maximal population of spermatozoa is established and constantly replenished within the fallopian tubes 12-24 hours after mating. The duration of survival of the spermatozoa in the uterus and fallopian tubes is approximately 10-12 hours. The rate and pattern of sperm transport is affected by the levels of circulating sex hormones and the effects that these hormones produce on the reproductive tract. The cervical mucus, which acts as a barrier to sperm during the majority of the cycle, becomes very fluid-like due to the action of oestrogens at oestrus and sperm penetration, the establishment of the cervical reservoir and subsequent transport are facilitated. This is achieved by a reorganisation of the mucus structure. The water content of the mucus increases due to the influence of oestrogen. The more fluid mucus develops a strain line effect which orientates the spermatozoa within the mucus strands. Excessive oestrogen can lead to a breakdown of these strands and as a consequence a disorientation of the spermatozoa and a failure of sperm transport.

The intrinsic movement of the sperm themselves (sperm motility) is crucial for the penetration of the cervical mucus and passage through the cervix.[20] Transport of spermatozoa along the remainder of the reproductive tract is aided by the peristaltic-like contractions of the tract.

There appears also to be a need for the spermatozoa to undergo some changes (capacitation) before they are capable of fertilisation and these changes require exposure to the tract environment for about one and a half to four hours. The ovum enters the ampulla of the fallopian tube and it is in this region that fertilisation takes place. The ovum is only capable of being fertilised for a relatively short period of time (twelve hours) once it enters the oviduct. As ovulation occurs about 25 - 30 hours after onset of oestrus in the ewe, spermatozoa numbers should be maximal at this time and fertilisation can occur almost immediately.

Fertilisation is achieved by the penetration of one spermatozoa through the zona pellucida of the ovum with the fusion of the nuclei of the spermatozoa and ovum to form the first cell of the zygote. Following the penetration of one sperm into the ovum, a reaction takes place in the zona which inhibits the penetration of other spermatozoa (polyspermy). The fertilised ovum descends through the oviduct and at the same time commences its cleavage divisions or segmentation.[26]

Embryo Development and Foetal Growth

Following fertilisation there is an initial rapid increase in cell numbers of the embryo with little increase in size. The two cell stage is reached about 30 hours after ovulation, the four cell by 36 hours, and eight cell by 51 hours. At this time the embryo has generally moved from the ampulla to the isthmus portion of the fallopian tube. The 16 cell stage is reached about 72 hours after ovulation. The embryo passes into the uterus at this stage (about 96 hours after ovulation). By 120 hours the majority of embryos have reached the 32 cell or morula stage and at eight days they reach the blastocyst stage and have shed the zona pellucida. This stage is characterised by a central cavity and presence of an embryonic disc. About the 12th day after mating the embryo commences to elongate extremely rapidly and within a period of 24 hours changes from a blastocyst of about 1 mm in diameter to an elongated sac 12 - 33 mm in length. This rapid growth continues and by day 14 - 15 has attained a length of 70 - 100 mm.[30]

Pregnancy

The presence of a viable embryo at day 13 is the signal for the maintenance of the corpus luteum and establishment of pregnancy. Removal of the foetus before this time leads to luteal regression and an oestrous cycle of normal length.[22]

From about the 15th day the embryonic membranes (amnion, allantois, chorion) are formed and by about the 25th day definite attachment of the cotyledons on the allanto-chorion and the caruncles on the uterine wall is apparent. This process of implantation is a very gradual one and in the early stages attachment is relatively tenuous and a period of some weeks may elapse before appreciable contact occurs between embryonic and maternal tissues.

During this period before the nutritional needs of the embryo can be met by the placenta, the embryo is nourished by the uterine secretions. Fig. 9.8 shows a diagrammatic representation of placentation in the

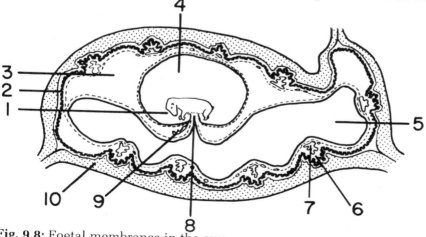

Fig. 9.8: Foetal membranes in the ewe

1. Foetus **2.** Allantochorion **3.** Extraembryonic coelom **4.** Amniotic cavity **5.** Allantoic cavity **6.** Cotyledon **7.** Foetal vessels **8.** Navel cord **9.** Yolk sac **10.** Uterus.

ewe. The weights of both the placental and foetal membranes increase rapidly in early pregnancy and maximum weight is achieved about half-way through pregnancy. A similar pattern is shown by the volume of foetal fluids.[29]

The foetus makes relatively little growth until the placenta is firmly established. Thereafter it grows at an increasing rate and doubles its weight in the last month of pregnancy. The rate of growth in the latter stages is influenced by the level of nutrition of the ewe. The duration of pregnancy in the ewe is approximately five months and birth occurs from day 144 to 150 depending on breed and litter size.

The corpus luteum remains active throughout pregnancy although during the last third of pregnancy the placenta is also producing progesterone and pregnancy can be maintained even if the ovaries are removed.

Although the actual trigger for parturition is not known, one of the first changes is an increase in the production of corticoid hormones by the foetal adrenal gland. This leads to an increase in the formation of oestrogens in the placenta. A rise in levels of total oestrogen in the maternal plasma takes place about day 100 of pregnancy and levels remain elevated for the last third of gestation. Immediately prior to parturition there is a fall in the levels of progesterone and this is accompanied by a very marked increase in levels of oestradiol and PGF.

The Birth Process

Parturition is normally divided into three stages.

Stage 1 is characterised by the dilation of the the cervix and contractions of the longitudinal and circular muscle fibres of the uterine wall and it generally lasts from between 1 to 24 hours.

Stage 2 or the expulsive stage is characterised by the complete dilation of the cervix, contraction of the abdominal diaphragmatic muscles, and final delivery of the foetus through the vulva. This stage has a duration of from 30 minutes to two hours.

Stage 3. This stage encompasses the expulsion of the placenta and is a complex process involving both mechanical and hormonal factors; it takes from 30 minutes to eight hours. Following parturition the involution of the uterus takes place and is usually complete at about 30 days after birth.

FACTORS AFFECTING THE REPRODUCTIVE PERFORMANCE OF SHEEP

Factors Affecting Semen Production and Characteristics

Season: Under New Zealand conditions it has been noted that a marked decline in volume, density and motility of semen occurs during the late spring and early summer period, coincident with the period of increasing day length, while maximum values are obtained in the autumn. Similar seasonal patterns have been reported elsewhere. Elevated temperature has a marked effect on semen quality under laboratory conditions, but there is little New Zealand evidence for any deleterious effects in the field. The level of humidity, and the amount of wool cover on the scrotum could influence the degree of any effect of temperature.

Frequency of ejaculation: The ejaculates of sexually active rams in the mating flock are considerably lower in volume and density than those collected from sexually rested animals.[1][31]

Nutrition: Prolonged undernutrition has led to a decrease in ejaculate volume, semen density, motility and survivability and also increases in the proportion of abnormal sperm.[23] In addition, the accessory sex glands may be influenced, with consequential changes in the composition and biological properties of the seminal plasma resulting from reduced testosterone secretion. The effects of nutrition are due mainly to level of energy intake. Quantity and quality of protein appear to be relatively unimportant unless it results in depressed food intakes.[6] Vitamin A deficiency can have a marked effect on spermatogenesis and adversely affects semen quality and fertility. Deficiency of various minerals such as zinc can also adversely affect semen quality.[34]

Age: The maximum rate of sperm production is attained shortly after puberty and further increases in the reserves of sperm in the testes are caused by increases in the size of the testes which increase with body weight and age up to about 3 years.

TABLE 9.1: Characteristics of Semen

	Ram	Bull	Boar	Stallion
Volume/ejaculate (ml)	1.0-1.5	4.0-8.0	150-250	50-120
Concentration of sperm (no $\times$ 10^6/ml)	2,000-6,000	1,500-2,500	150-300	150-300
Total sperm per ejaculate ($\times$ 10^9)	2.0-9.0	6.0-20.0	22.5-75.0	7.5-36.0

Breed: Differences between breeds of ram have been reported in terms of semen production and testicular reserves but it is possible that this is a reflection of breed differences in body weight. There are few New Zealand data on this aspect.

Disease: Various diseases such as brucellosis and scrotal mange can affect the quality and quantity of the semen.

Factors Affecting Oestrous Behaviour

Day length. The major factor influencing the seasonal pattern of oestrous activity seen in sheep is day length. The majority of non-tropical breeds exhibit oestrus during the period of decreasing daylight hours (autumn-winter). However some exhibit oestrus in the spring months.

Nutrition. There is little evidence to suggest any effect of level of nutrition on both the duration of oestrus and the breeding season. However under conditions of severe undernutrition oestrous activity may cease and the animals become anoestrus.

Age. The duration of oestrus is shorter and the intensity of oestrous behaviour less pronounced in hoggets and maiden two-tooth ewes than it is in older ewes. As a result the younger ewes have a shorter time in which to mate and may not compete with older ewes under a system where ewes of different ages are mated together.

Younger ewes also have a shorter breeding season than do older ewes and this is due to an earlier cessation of reproductive activity.[13] It has been reported, however, that fewer two years old ewes than mature ewes were mated in the first two weeks of the joining period.[15]

Age, Liveweight and Puberty: Puberty is more closely related to live weight than to age, with ewes growing at a faster rate exhibiting their first oestrus earlier and at a heavier body weight than those growing slower. However, both the live weight and age of puberty can be markedly affected by the genotype or breed of the ewe lamb. Lambs of the more fecund breeds e.g. (Finnish Landrace) tend to reach puberty at an earlier age and lower body weight than those of the less fecund breeds.[8] The season of birth in relation to the next breeding season can influence the age of puberty in that later born lambs may not be old enough nor heavy enough to exhibit oestrus in their first breeding season. Also early born lambs tend to be older and heavier at time of first oestrus due to the effect of season or day length. The birth rank of the lamb also has an influence with twin ewes tending to exhibit first oestrus at a lower body weight but at a similar age.[12]

Breed or genotype. The length of the breeding season is markedly affected by breed. Under New Zealand conditions breeds such as the Dorset Horn have an earlier onset of oestrous activity than do the Coopworth, Perendale or Romney. The length of breeding season is longer in the Dorset than in the Coopworth which in turn is longer than the Romney with the Perendale intermediate between these.[13] There is also evidence that the duration of oestrus may be influenced by genotype with the more fecund breeds having a slightly longer duration of oestrus and in particular a longer interval from oestrus to ovulation.

Rams. The presence of the ram can have a marked influence on the duration of oestrus. Ewes in constant contact with rams have a shorter duration than those which are teased intermitttently. In addition, the introduction of rams into a flock prior to the normal breeding season, can stimulate and sometimes synchronise the early onset of oestrus (i.e. ram effect). There appears to be an effect of breed of ram on the magnitude and efficiency of this effect with Dorset rams being superior to Romney rams.[33]

Lactation. The presence of a suckling lamb can prolong the period post-partum that a ewe remains in anoestrus and early weaning can lead to a more rapid resumption of oestrous activity. This is, however, confounded by breed and seasonal interactions. In New Zealand this does not have any great effect as the majority of ewes lamb during the seasonal anoestrous period and are weaned before the breeding season resumes.

Location. Within New Zealand it appears that the latitude at which the ewes are kept can influence the time of onset of the breeding season with a tendency for an earlier but less precise onset in the northern regions.

Factors Affecting Ovulation Rate

Live weight and live weight change. It is well established that there is a positive relationship between live weight and ovulation rate. (see Chapter 10).

Nutrition. It has been generally considered that energy level rather than protein level of flushing feed is the most critical factor in influencing ovulation rate.[27] Other components of the diet also can be important. Marked increases in ovulation rate have been obtained with as little as seven days of feeding on lupin grain in the absence of any live weight changes. These increases have not been found when other grains were fed.[17] This response to lupins appears to be modified however, by the type of diet pre-treatment with the most marked response in ewes

231

grazing dry stubble and least response in ewes grazing green pasture. The difference in live weight-ovulation rate relationship between ewes flushed on pasture and ewes flushed on silage is also of interest in this connection.[28]

There are other effects of nutrition on ovulation rate and the most important of these would appear to be the depression in ovulation rate observed in ewes flushed on either oestrogenic clovers (Red clover and Subterranean clover containing formononetin) or oestrogenic lucerne (containing coumestans). Ewes flushed on oestrogenic lucerne have shown a depression of 30% in ovulation rate compared to ewes of similar weight that have been flushed on grass pasture. However, this effect is a temporary one and can be overcome by a short period (14 days) of grass feeding prior to joining.

Genotype. The breed of ewes and strain or selection line within a breed can markedly influence ovulation rate. The Finn and Romanov are examples of breeds with very high ovulation rates. The Border Leicester and Dorset Horn and their crosses have a higher ovulation rate than the Romney. Selection within a breed can increase ovulation rate (e.g. Booroola Merino and the high fertility Romneys).

Season and stage of the breeding season. There appears to be a trend for the ovulation rate to be highest in mid-autumn (April-May).[2] The ovulation rate increases rapidly to this peak and then gradually declines. The considerable variability in reports on effect of cycle number on ovulation rate is probably dependent on the date of first oestrus. If it is early (February-March) then ovulation rates will initially increase with cycle number but if first oestrus is later (April) then ovulation rate will commence at the peak and decline with succeeding cycles. The nutritional regime of the animals during this period is a major complicating factor.

Age. The ovulation rate is lower in hoggets, two-tooth and very old ewes than it is in three to five year old ewes. In the case of the two-tooth ewes this may be an effect of body weight as well as that of age.

Factors Affecting Fertilisation

Semen quality and quantity. This is the most obvious factor that can limit the level of fertility in a flock and the factors influencing this have been presented previously. However, with the exception of the extremes, the relationships between semen measurements and fertility in the field are generally low.

There are some recent suggestions that the genotype of the ram may influence fertilisation rate with rams of the more fecund breeds or strains having a higher fertility, although this could be a libido effect as well as a semen effect.

Mating behaviour of the ram and ewe. There is a linear increase in conception rate with increases in number of times a ewe is mated in one oestrous period and also with the number of rams to which she mates.[16] Thus the mating dexterity and libido of the ram and the intensity and length of oestrus could influence the fertilisation rate.

The time of mating relative to the onset of oestrus. This is generally only of consequence when artificial insemination or hand mating is being used. Single matings very early in oestrus or later than 20 hours after the onset tend to have reduced levels of fertilisation.

Nutrition. There is a large mass of very variable and often contradictory data on the effect of level of nutrition on fertilisation rates.[27] In general it appears that only very low levels of nutrition may reduce fertility and this effect is confounded with the effects of level of nutrition on oestrous behaviour and ovulation rate. Some specific nutritional factors can influence fertility. Deficiencies in copper, manganese, zinc and selenium can reduce fertilisation. Oestrogenic legumes particularly Subterranean and Red clovers have been associated with marked reductions in conception rate, the degree of effect being related to the concentration of the isoflavone formononetin.

Hormone treatment. Manipulation of the normal endocrine pattern in the ewe by use of oestrous synchronisation treatments have often been shown to cause a reduction in fertility.

Factors Affecting Embryo Survival

The failure of the fertilised ovum to develop, implant and progress through pregnancy and result in a lamb is a major source of reproductive wastage. This, on average, occurs with about 20 - 30% of the ova shed. The major period of loss is in the first 30 days of pregnancy with most occurring just prior to day 18. The age of the embryo when loss occurs affects the duration to return to oestrus. If the complete uterine complement of embryos is lost then embryo death prior to day 13 has no effect on cycle length. Loss after day 13 causes a delay in the return to oestrus. The magnitude of the delay is proportional to age of the embryo when loss occurs although there is considerable variability.

The factors contributing to this pre-natal mortality have been reviewed.[10]

Ovulation rate. There is clear evidence to indicate that embryonic mortality increases with increasing ovulation rate. This reaches a level where complete loss of all embryos may occur especially under influence of super-ovulation treatments. However some recent data indicate that in those ewes with two ovulations, the proportion of ewes that have only one lamb decreases as the ovulation rate of the flock increases, suggesting that there is a negative relationship between flock fecundity and embryonic mortality.

Breed. There is little information on differences between breeds, and what differences do exist may be confounded with differences in ovulation rate.

Age. It appears that at extremes of age, hoggets (9 months) and very old ewes (8+ yrs), have higher levels of embryonic loss than do other age groups.

Nutrition. The effect of nutrition on embryonic loss is not clear. The effects of pre-mating nutrition are confounded by their effects on ovulation rate. The effects of post-mating nutrition indicate that only very severe undernutrition in early pregnancy can produce an increase in embryonic loss. Certain mineral deficiencies (e.g. selenium) may result in embryonic death.

Temperature. The effect of high temperatures on embryonic loss is well documented. Elevated temperatures early in pregnancy cause an increase in embryonic loss. The early cleavage stages of the embryo are the most susceptible. While most of the information has been obtained under controlled climate conditions, some evidence from field observations has been reported but it is unlikely that this constitutes a major source of loss under New Zealand conditions.

General stress. Some evidence exists that stress from causes other than high temperature and under-nutrition may be a contributing factor (e.g. shearing).

Season. The stage of the breeding season may have an effect with higher levels of embryonic death in the early and very late stages.

Disease. A number of diseases may terminate pregnancy in the sheep. They usually cause abortion and have their main impact late in pregnancy. While the incidence of such disease-related losses may be severe in individual flocks their total economic impact is generally not great.

Factors Affecting Losses About Birth

The major causes of neonatal loss are dystocia (difficult birth) and exposure and starvation, with disease and other non-specific factors generally being only sporadic in their effects.

234

The incidence of loss is influenced by the breed of ewe, with the Romney showing a higher loss than the Perendale or Coopworth. Within a breed the incidence can be markedly influenced by individual sires.[11]

The birth rank of the lamb has a major influence on both the incidence and cause of loss. The loss rate is higher in twins and other multiple births and these tend to suffer more from exposure and starvation. In single-born lambs the major cause of loss is dystocia.

Nutrition of ewes during late pregnancy can have a marked influence on neonatal loss. Underfeeding has led to light lambs with lower fat deposits and less liver glycogen and because of their limited energy reserves and the relatively high surface area to body mass, they are more prone to loss due to exposure. They may also suffer from a loss of suckling drive and this causes physiological starvation.[27] Energy deficiency in late pregnancy can result in pregnancy toxaemia with the loss of both ewe and lamb. Excessively high levels of feeding may produce large foetuses and increased losses due to dystocia, particularly in singles.

Certain specific nutritional factors such as high pasture oestrogens can cause dystocia through the development of uterine inertia.

The major cause of dystocia is disproportionate foetal size in relation to the pelvic area of the ewe. The highest incidence of loss is with the largest lambs and in younger and smaller ewes.

The main factors contributing to loss through exposure and starvation are the climatic conditions at the time of lambing. Low temperatures and wind with rain or snow result in excessive heat loss from the lamb. Low birth weights predispose lambs to loss from this cause, and poor mothering ability of the ewe can be a contributing factor.

CONCLUSIONS

The outline of the mechanisms controlling the reproductive process in sheep highlights those stages of the process that are most susceptible to change. The factors most likely to affect these stages and therefore influence the overall reproductive performance are nutrition, season, breed and age. Therefore management decisions relating to these factors are likely to have effects on the reproductive performance of a flock and thus on the profitability of the enterprise.

REFERENCES

[1] Allison, A. J. 1972: The effect of mating pressure on characteristics of the ejaculate in rams and on reproductive performance in ewes. *Proc. N.Z. Soc. Anim. Prod. 32:* 112.

[2] Averill, R. L. W. 1965: Variability in ovulatory activity in mature Romney ewes in New Zealand. *Wld. Rev. Anim. Prod. 3:* 51

[3] Baker, T. G. 1972: Primordial germ cells. In *Reproduction in Mammals* 1. *Germ cells and fertilization.* (Eds C. R. Austin and R. V. Short) Cambridge University Press.

[4] Baker, T. G. 1972: Oogenesis and ovulation. In *Reproduction in Mammals.* 1. *Germ cells and Fertilization.* (Eds C. R. Austin and R. V. Short) Cambridge University Press.

[5] Bassett, E. G. 1965: The anatomy of the pelvic and perineal regions of the ewe. *Aust. J. Zool. 13:* 201.

[6] Braden, A. W. H.; Turnbull, K. E.; Mattner, P. E.; Moule, G. R. 1974: Effect of protein and energy content of the diet on the rate of sperm production in rams. *Aust. J. Biol. Sci. 27:* 67.

[7] Courot, M. 1967: Endocrine control of the supporting and germ cells of the impuberal testis. *J. Reprod. Fert., Suppl. 2:* 89.

[8] Dyrmundsson, O. R. 1973a: Puberty and early reproductive performance in sheep. I. Ewe lambs. *Anim. Breed. Abstr. 41:* 273.

[9] Dyrmundsson, O. R. 1973b: Puberty and early reproductive performance in sheep. II Ram lambs. *Anim. Breed. Abstr. 4:* 419.

[10] Edey, T. N. 1969: Prenatal mortality in sheep: A review. *Anim. Breed. Abstr. 37:* 173.

[11] Hight, G. K.; Bigham, M. L. 1975: Aspects of lamb mortality in New Zealand flocks. *Proc. Workshop Perinatal Mortality in Farm Animals. CSIRO.* (Ed. G. Alexander.)

[12] Hight, G. K.; Lang, D. R.; Jury, K. E. 1973: Hill country sheep production. V. Occurrence of oestrus and ovulation rate of Romney and Border Leicester × Romney ewe hoggets. *NZ Jl Agric. Res. 16:* 509.

[13] Kelly, R. W.; Allison, A. J.; Shackwell, G. H. 1976: Seasonal variation in oestrous and ovarian activity of five breeds of ewes in Otago. *NZ Jl Exp. Agric. 4:* 209

[14] Knight, T. W. 1977: Methods for the indirect estimation of testes weight and sperm numbers in Merino and Romney rams. *NZ Jl Agric. Res., 20:* 291.

[15] Knight, T. W.; Hight, G. K. 1976: Variations in the fertility over the joining period of ewes on hill country. *NZ Jl Agric. Res. 19:* 211.

[16] Knight, T. W.; Lindsay, D. R. 1973: Identifying the mating performance of individual rams in field flocks. *Aust. J. Agric. Res., 24:* 579.

[17] Lindsay, D. R. 1976: The usefulness to the animal producer of research findings in nutrition on reproduction. *Proc. Aust. Soc. Anim. Prod. 11:* 217.

[18] McNatty, K. P. 1978: Follicular Fluid. In *Evaluation of the Vertebrate Ovary.* (Ed. R. Jones) Plenum Press.

[19] Mann, T. 1959: Biochemistry of semen and secretions of male accessory organs. In *Reproduction in Domestic Animals Vol. 2.* (Eds H. H. Cole, and P. T. Cupps) p 51.

[20] Mattner, P. E. 1966: Formation and retention of the spermatozoan reservoir in the cervix of the ruminant. *Nature, Lond., 212;* 1479.

[21] Mauleon, P. 1967: Oogenesis and folliculogenesis. In *Reproduction in Domestic Animals* 2nd ed. Chapter 7. (Eds H. H. Cole, and P. T. Cupps) Academic Press.

[22] Moor, R. M.; Rowson, L. E. A. 1966: The corpus luteum of the sheep: Effect of the removal of embryos on luteal function. *J. Endocr. 34:* 497.

[23] Moule, G. 1970: Australian research into reproduction in the ram. *Anim. Breed. Abstr., 38:* 185.

[24] Moule, G. R.; Knapp, B. 1950: Observations on intra-testicular temperatures of Merino rams. *Aust. J. Agric. Res., 1:* 456.

[25] Ortavont, R. 1959: Spermatogenesis and morphology of the spermatozoon. In *Reproduction in Domestic Animals. Vol II Chapter 1* (Eds. H. H. Cole; P. T. Cupps) Academic Press.

[26] Perry, J. S. 1971: *The ovarian cycle of mammals.* Oliver and Boyd Edinburgh.

[27] Rattray, P. V. 1977: Nutrition and reproductive efficiency: *Reproduction in domestic animals.* 3rd Ed. Chapter 21. (Eds H. H. Cole, P. T. Cupps). Academic Press.

[28] Rattray, P.V.; Jagusch, K. T.; Smith, J. F.; Tervit, H. R. 1978: Flushing ewes on pasture and pasture silage. *Proc. NZ Soc. Anim. Prod., 38:* 101.

[29] Robinson, T. J. 1957: Pregnancy. In *Progress in the Physiology of farm animals.* (Ed. J. Hammond). *Butterworths Lond. 3:* 793.

[30] Rowson, L. E. A.; Moor, R. M. 1966: Development of the sheep conceptus during the first fourteen days. *J. Anat. 100:* 777.

[31] Salamon, S. 1964: The effect of frequent ejaculation in the ram on some semen characteristics. *Aust. J. Agric. Res., 6:* 950.

[32] Salamon, S. 1976: *Artificial Insemination of Sheep.* Publicity Press - Sydney.

[33] Tervit, H. R.; Havik, P. G.; Smith, J. F. 1977: Effect of breed of ram on the onset of the breeding season in Romney ewes: *Proc. NZ Soc. Anim. Prod., 37:* 142.

[34] Underwood, E. J.; Somers, M. 1969: Studies of zinc nutrition in sheep. 1. The relation of zinc to growth, testicular development, and spermatogenesis in young rams. *Aust. J. Agric Res., 20:* 889.

[35] Waites, G. M. H.; Setchell, B. P. 1969: Physiology of the testis, epididymis and scrotum. In *Advances in Reproductive Physiology* (Ed. A. McLaren).

[36] Watson, R. H.; Sapsford, C. S.; McCance, I. 1956: The development of the testis, epididymis and penis in the young Merino ram. *Aust. J. Agric Res., 7:* 574.

10
Techniques of Modifying Reproductive Performance

A. J. Allison
Invermay Agricultural Research Centre, Mosgiel
Consultants: *T. W. Knight* and *M. F. McDonald*

SUMMARY

Reproductive activity of sheep in various physiological conditions can be altered through the use of a range of techniques. These can be applied to regulate the timing of reproduction (control of oestrus, ovulation, parturition), to increase the number of offspring produced per pregnancy (control of ovulation rate, litter size), or to extend the influence of selected individuals in the flock (artificial insemination, ova transplantation). The application of such techniques in the New Zealand sheep industry is briefly discussed.

INTRODUCTION

Many managerial and hormonal techniques for modifying reproductive performance could be adopted to increase the efficiency of reproduction. Lamb tailing percentages in the New Zealand flock have varied between 92 and 100% in the last 20 years and there must be much scope for improvement. Increases in efficiency of reproduction may come from more lambs per pregnancy, more lambings per ewe lifetime, an increase in the rate of lamb survival or in a greater weight of lamb weaned per ewe. Available techniques will be discussed with some comment as to their application in the New Zealand industry.

INDUCTION AND SYNCHRONISATION OF OESTRUS

Hormonal synchronisation of oestrus (heat) includes three main categories. Firstly, immature ewe hoggets may be stimulated to ovulate prior to normal puberty, lactating or anoestrous ewes may be stimulated to ovulate in the non-breeding season or ewes may be treated prior to the onset of the breeding season to advance the occurrence of oestrus. Secondly, within the breeding season ovulation and oestrous activity may be suppressed by administration of progesterone or synthetic analogues (*progestagens*), and following the treatment which outlasts the *luteal phase* of the cycle the ewes will display oestrus and

239

ovulate. Thirdly, synchronisation may occur following the regression of corpora lutea at a predetermined time. This is done by injecting ewes with a prostaglandin analogue which acts in the same way as the release of prostaglandin from the uterus causing luteal regression and termination of the cycle.

Administration of progestagens may be by addition to feedstuffs, by injection, and by insertion of intravaginal sponges or implants impregnated with the hormone. Injection of progesterone daily or every two days is tedious and inconvenient. The pattern of onset of oestrus and control of ovulation following the use of oral progestagens is imprecise in comparison with injections, but good results have followed the use of intravaginal sponges or implants which offer a simple and effective means of continuous administration. Many progestagens have been tested[45] for their ability to control time of oestrus and ovulation; some are highly potent with a short duration of activity resulting in precise synchronisation of oestrus upon cessation of treatment. The use of intravaginal sponges provides a simple method to control oestrus but there may be problems in maiden ewes due to the difficulty of insertion past a vaginal constriction. The use of narrow diameter sponges inserted through a smaller applicator will avoid such difficulty.

Induction of Oestrus in Ewe Hoggets

Hormonal treatments to induce breeding in young sheep have been reviewed.[23] Providing animals are 5-7 months of age, progestagen given for 12-14 days followed by 600-800 iu of Pregnant Mare's Serum Gonadotrophin (PMSG) will result in oestrus and ovulation in 70-100% of hoggets. Conception rates achieved have varied from 40-80%. More success may be expected the closer the ewe hoggets are to the onset of their normal breeding season although animals induced to ovulate will not necessarily continue cyclic activity. There is little information concerning lambing performance of hoggets following mating at an induced heat, but the use of these techniques would have little relevance in the New Zealand industry. Rather, more emphasis should be given to improving nutrition which would result in better grown animals with a high incidence of oestrus occurring naturally and usually about one month after the normal breeding season in mature ewes.

Induction of Oestrus in Mature Ewes in the Non-breeding Season

Most ewes are anoestrus from the time of lambing in August-September until the following February-March. As virtually all mature ewes are joined early in the breeding season induction of oestrus, if attempted, will either be in lactating ewes or soon after weaning depending on the desired time of hormonally induced lambing. The principles involved in the induction of oestrus and ovulation in anoestrous ewes apply also to lactating ewes but there are additional problems early in lactation.[27] [35] A period of progestagen treatment followed by a single injection of PMSG will result in oestrus and ovulation in a variable proportion of ewes but fertility has often been poor.

During pregnancy the uterus expands greatly and following lambing it returns to its original form (*involution*) within four weeks. Blood and cellular debris may still be present for a further two weeks. Extremely low fertility in ewes induced to breed early in lactation is probably attributable to faults within the tract.

The viability of ova of lactating ewes may be less than those recovered at the same time of year from ewes which had not lambed.[35] Failure to conceive at induced heat shortly after lambing may also be partially due to the absence of a preliminary ovulation.[27] The best responses for oestrus and fertility are usually obtained using two injections of PMSG a cycle length apart with an intermediate period of progestagen treatment. Increased fertility is usually obtained when oestrus is induced later in lactation.[25]

Limited information exists on the interval for ewes to return to cyclic activity after lambing within the breeding season (the result of mating while in anoestrus). In considering breeding programmes for lambing at intervals less than a year the time necessary for involution of the uterus and for silent ovulation to occur must be taken into account. Consequently, breeding routinely at six-monthly intervals appears impractical. However, breeding of ewes three times in two years may be feasible using two matings within the breeding season (i.e. July - lambing December and March - lambing August, followed by a hormonally induced mating in November to lamb in April). Such a programme would need careful consideration before commercial application. Feed requirements for lactating ewes at different times of the year and the cost of such feed could be the main limitations to the breeding of ewes at intervals of less than a year. Breeds such as the Merino, Poll Dorset and Dorset Horn which display longer breeding seasons than the Romney may be of advantage. More information is also sought on the factors of management such as nutrition and time of weaning that might extend the breeding period of sheep.

The libido and semen quality of rams may decline during the spring and summer in contrast to the autumn. Post-partum heats may be short and of low intensity, making sexual activity of rams critical. Low fertility in matings in spring and summer is therefore likely.

Induction of Earlier Breeding Activity by Treatment with Progestagens

Administration of progestagens to mature ewes just prior to the expected onset of the breeding season causes most to display oestrus after cessation of treatment. Fertility will usually be lower than at normal oestrus but ewes which do not conceive will return to service 16-17 days later whereas, following mating at an induced heat in the middle of the non-breeding season, ewes will not return to oestrus. This technique is usually more effective than induction and synchronisation of oestrus that can be caused by a "ram effect" although more costly and labour intensive.

Synchronisation of Oestrus Within the Breeding Season

Progestagens

The time of oestrus and ovulation within the breeding season may be controlled following the luteal phase of the cycle being prolonged by the administration of a progestagen. Synchronisation techniques have not been used widely in commercial sheep production, except in France and Ireland. In those countries the techniques have been used mainly in conjunction with PMSG to increase lamb drop and also to breed ewes more frequently.

The use of intravaginal sponges containing progestagen provides the most convenient method of oestrus control in ewes. The level of hormone and the period of administration must be sufficient to suppress oestrus during treatment. Following sponge withdrawal, oestrus usually occurs within three days. The interval from withdrawal to oestrus depends on several factors including:

(1) The dose level of progestagen;[9]
(2) Method of preparation of the sponge;[46]
(3) Whether or not PMSG is injected at the end of progestagen treatment. A typical pattern of results for oestrus and lambing in ewes after sponge treatment is shown in Fig. 10.1.

The fertility of ewes inseminated or mated following progestagen synchronisation of oestrus is variable and is usually lower than in untreated ewes, irrespective of the method of administration. The depression in fertility has ranged from less than 10% to more than 50% and appears unpredictable. The lower fertility is due to impaired transport and survival of sperm within the reproductive tract. Fertility at the second oestrus following sponge withdrawal is normal although synchronisation is less precise.

Refinements in mating management and techniques of artificial insemination have resulted in fertility often comparable to that in untreated ewes. The first report of synchronisation of oestrus with intravaginal sponges showed that high fertility was achieved in ewes inseminated with 0.2 ml of undiluted semen or joined with a high

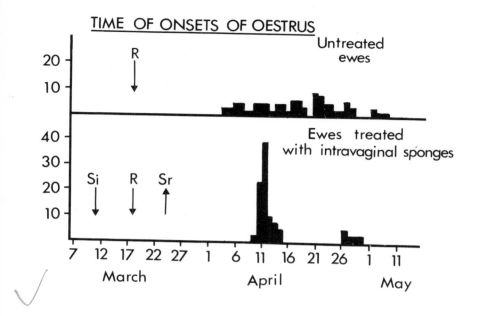

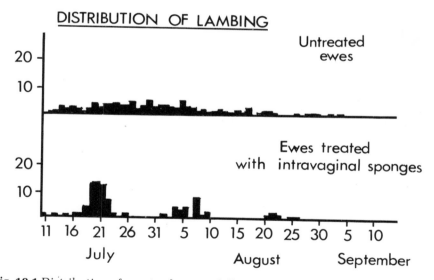

Fig. 10.1 Distribution of onsets of oestrus following withdrawal of intravaginal sponges and the distribution of subsequent lambing
(Se = sponges inserted; Sr = sponges removed; R = rams joined) From Clarke et al. 1966[16]

percentage of rams.[44] Density of the inseminate is critical in achieving acceptable fertility in progestagen-treated ewes.[10] Ewes should be inseminated with undiluted semen if possible. In controlled breeding programmes using natural service 10 ewes per ram or less is advocated and joining rams with ewes 48 hours after sponge withdrawal, rather than at the time of withdrawal, may substantially improve lambing rates.[28] This is probably due to more ewes being mated at an optimal time, and a more dense inseminate being deposited in the anterior vagina.

The detrimental effect of stress on the transport of spermatozoa within the reproductive tract may often have been underestimated. In synchronised ewes where the transport of sperm is impaired, factors which may further impair this could be expected to decrease fertility. Consequently it is advisable to treat ewes carefully during artificial insemination. Various workers have suggested that the size of the group awaiting insemination should not exceed 40-50, presumably due to greater agitation in larger flocks.

There are few managerial advantages of synchronisation although there is the possibility of greater efficiency in use of feed at critical times. The spread of lambing will not be reduced in comparison with a normal mating programme but will occur in discrete peaks (see Fig. 10.1). If large numbers of ewes are required in oestrus at a particular time for artificial insemination then this may be effectively achieved with progestagen administration, but it is best to inseminate at the second oestrus following the end of treatment to obtain normal fertility. Insemination or mating will therefore be 17 days later which may be a disadvantage in commercial flocks.

Prostaglandins

A single injection of prostaglandin on days 5 to 15 of the cycle will cause a similar pattern of synchronised oestrus to that observed following withdrawal of intravaginal sponges. If two injections are given 10 days apart then, at the time of the second treatment, all ewes are likely to have a *corpus luteum* which will later regress allowing oestrus to occur. Prostaglandins are of no use early in the breeding season or for out-of-season breeding as an active corpus luteum is absent. Some depression in fertility occurs in prostaglandin-treated compared with untreated ewes.[8] Two injections rather than one appear necessary for good results. As with intravaginal-sponge treatment, fertility at the second oestrus after injection is normal although synchronisation is less precise.

Effect of Introduction of Rams on the Pattern of Onset of Oestrus

Prior to the onset of the breeding season most ewes will ovulate without displaying oestrus. This is commonly known as a *'silent'* heat. Joining rams with ewes close to the start of the breeding season but just before the first silent heats occur will stimulate many to ovulate within

3 to 6 days and display their first oestrus 17 days later. Romney ewes therefore may show a pronounced peak of oestrus 3 weeks after introduction of rams resulting in an average date for first oestrus 10 days earlier than is usual for the region.[24]

The timing of introduction of rams is critical as rams joined too early have no effect on the pattern of onset of oestrus and rams introduced later can affect only those ewes which have not already experienced a silent heat.

In Northern regions of New Zealand the time of ram introduction likely to achieve greatest synchronisation appears to be at the beginning of February but it will be later in Southern areas owing to a later onset of the breeding season. Results of the technique are not always predictable as the onset of the breeding season may vary by as much as two weeks between years on the same farm, and also between farms in the same area in the same year. Entire rams are sometimes considered to be better for synchronising oestrus than are vasectomised rams, but if so the reason is unknown.

"Ram synchronisation" may be of advantage if an earlier lambing date is desired, as when summer drought or cropping requirements make finishing of lambs more difficult or when a price premium exists for earlier marketed animals. Also if fertile mating of ewes is delayed until second oestrus after ram synchronization then the ovulation rate, and therefore the potential for twinning, may be higher.[1]

Some breeds of ram such as the Dorset may cause earlier onset of oestrus in ewes than occurs with Romney rams[49] but the effect is small and of limited practical significance.

METHODS OF INCREASING OVULATION RATE

Nutrition and Live Weight: Flushing

The relationship between pre-mating live weight, barrenness and twinning in sheep is depicted in Fig. 10.2.[19] This type of relationship is evident among several breeds in New Zealand, and the between flock effects may be even greater than the observed 6% increase in twinning per 4.5 kg increase in live weight. Barrenness increases markedly below 40 kg live weight although in Merino ewes given sound management at mating high fertility is possible.[6][7] Twinning increases linearly with increasing live weight, at least up to 70 kg, and there is no evidence of a decline at the highest weights. Consequently improved nutrition, which results in higher live weights, is one of the most practical methods of increasing reproductive rates available to commercial breeders.

Improving nutrition prior to mating (*flushing*) is widely advocated and adopted. For best results it has been suggested that ewes should not be too high in condition at the start of flushing. However, in ewes

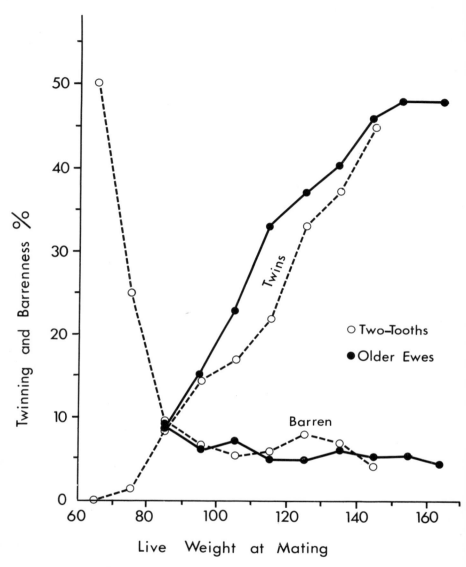

Fig. 10.2 Influence of live weight at mating on percentage barrenness and twinning in two-tooth and older ewes. From Coop 1962.[19]

deliberately made and kept as heavy as possible after weaning, compared to those which lost weight, a superior reproductive performance occurred even though some of the live weight advantage was lost and live weights converged when the groups grazed together during flushing. This superior performance was due to an increase in
246

ovulation rate.[51] Consequently there is no necessity to take weight off ewes in order to achieve a flushing response.

Flushing is a relative term. With the exception of feeding ewes on oestrogenic pastures, improved nutrition even when ewes may not be gaining live weight, will still result in higher twinning rates. The flushing response[20] has been shown to include:

(1) A *static effect* due to the increased live weight and not related specifically to the time of increased nutrition.

(2) A *dynamic effect* due to the change in live weight or condition brought about by the improved feeding.

These two effects contribute approximately equally to the flushing response. Increases in twinning due to the static and dynamic effects of flushing are related to increases in ovulation rate.[41] Flushing does not affect the pattern of onset of oestrus and has little effect on barrenness. However, the reduction of ewe live weights before flushing under the impression that only ewes in lean condition will respond to flushing can lead to a small increase in barrenness. Increases in twinning occur as the duration of flushing increases. In summary it seems that a 15-20% response in lambing percentage results from flushing for three weeks before mating and for another three weeks during mating. About half of this can be achieved by increasing weight at other times of the year. Also the difference in nutrition between flushed and not flushed groups in experimental conditions would be more marked than under usual farming conditions. Thus the advantages of flushing must be considered in relation to costs of supplying extra feed prior to mating. The most important principle is that heavier ewes will produce more lambs and that additional increases may be achieved by offering an increased plane of nutrition prior to and during mating.

It is important not to feed ewes on *lucerne* close to mating. Various studies have shown that ewes mated on lucerne may have considerably fewer twins than ewes mated on grass.[29] Slight increases in barrenness may also occur. *Flushing on red clover* will cause a reduction in lambing percentage[15] but little information exists as to the proportion of clover required in a sward before noticeable effects occur. It seems unwise to graze any breeding stock on pastures containing significant amounts of red clover, if this can be avoided, as permanent reductions in ewe fertility may occur.

Time of Mating

The percentage of ewes ovulating and the ovulation rate in ewes varies during the season. The mean time of onset of ovulatory activity begins earlier in the North Island of New Zealand than in the South and there is a marked seasonal change in ovulation rate with peak values in April and the first half of May.[12] In ewes which have limited live weight change over the mating period there is a rise in ovulation rate from the first to the third oestrus with a decline thereafter.[1][37]

A prolonged delay in the time of mating as when drought conditions prevail (or as possible in out-of-season breeding programmes) may result in lower ovulation rates and therefore lambing percentage. If live weight falls prior to the time of delayed mating then lambing performance may be further reduced. As the time of mating is delayed from April to June ewe barrenness is also increased. However, it may be good practice following drought, to delay mating by a few weeks so that the effects of improved nutrition may be obtained and cause an increase in live weight and lambing percentage.

Use of Gonadotrophins

The number of ovulations that occur when the ewe is mated may be increased by injecting ewes with follicle stimulating hormone (FSH) between the 12th and 14th days of the oestrous cycle. The most readily available source of FSH is in the blood serum of pregnant mares (PMSG) with maximum concentrations occurring about 2-3 months of pregnancy. Extracts from horse anterior pituitaries (HAP) also will increase the number of ovulations but the material is not readily available and has the disadvantage of having to be given in a series of injections over 2-3 days. At higher doses HAP may give a more consistent response than does PMSG.

As PMSG has to be injected at a specific stage of the oestrous cycle it is usually used in conjunction with or after detection of synchronised oestrus. Ovulation rates sometimes are higher if PMSG is administered several days before the end of progestagen treatment although, for convenience, injections are usually given at the time of withdrawal of intravaginal sponges containing progestagen. Ovulation rates obtained are linearly related to dose of PMSG (see Table 10.1) but at the highest doses many cystic follicles and low fertilisation rates occur. Ovulation rates at the subsequent oestrus are not affected. Trials in New Zealand have shown an average increase of 0.7 lambs born per ewe lambing following injection of 1000 i.u. PMSG.[52]

TABLE 10.1: Ovulation Responses to Injections of PMSG

Criterion	Dose of PMSG (iu)				
	0	250	500	1000	2000
Mean number of ovulations[50]	1.17	1.50	2.07	4.33	
Range	1-2	1-2	1-3	1-13	
Mean number of ovulations[43]			4.1	10.6	15.8
Range			2-9	4-33	8-29

Increases in litter size after the use of super-ovulation therapy are limited by the capacity of ewes to carry extra foetuses. This principle is illustrated in Fig. 10.3. Considerable embryonic mortality occurs before attachment of the blastocysts and this is followed by further mortalities up to about 3 weeks after mating, after which time there is little further loss.[43]

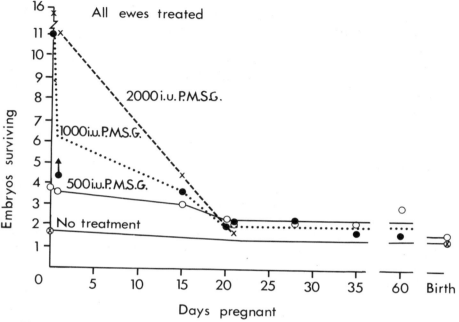

Fig. 10.3 Survival of embryos in relation to level of PMSG injection — ewes conceiving. From Robinson 1951.[43]

The greatest technical problem in using PMSG to increase lamb drop is the variation between ewes, between flocks of ewes, and between times of the year in the ovarian response to a constant dose. Also different batches of PMSG often give inconsistent responses due to non-standardization of potency of individual preparations. Ewes of higher natural fecundity will have considerably higher ovulation rates than ewes of lower fecundity[13] and ewes whose live weight has been limited by inadequate nutrition will not respond as readily as similar animals of heavier live weight.[2] If ewes are repeatedly injected with PMSG, as in an egg transfer programme involving several donor operations, then there is usually a decrease in the percentage of ewes ovulating and in ovulation rate.

In addition to uses in egg transfer programmes the main role for PMSG in the New Zealand industry is in flocks of low natural fecundity. It is essential that most of the increased lambs so obtained are reared to sale otherwise the initial advantage will be diminished.

The use of these techniques should only be regarded as an interim measure until naturally fecund animals, which will produce a much higher lambing performance under 'normal' management, can be bred and widely used within industry.

Breeding

The use of sires from highly fecund breeds over commercial ewe flocks is perhaps the easiest method of increasing ovulation rates in ewes. First cross progeny of sires such as the Border Leicester, and Booroola Merino have shown substantial increases in ovulation rates and lamb drop in comparison with straightbred ewes of the parent breeds.[1][11] The contribution of hybrid vigour, if any, to these increases must be assessed.

OVA TRANSFERS

Techniques

Large numbers of ova resulting from multiple ovulation (after PMSG injection) can be transferred to recipient ewes to increase the number of offspring from selected animals. The technique requires the recovery of fertilised ova from the Fallopian tubes or uterus of the donor ewes and transfer of these to the uterus of recipient ewes which are at a similar stage of the cycle.

An example of the sequence of events in the organisation of an ova transfer programme using progestagen synchronisation is in Table 10.2. Prostaglandins also might be used to synchronise the cycles of the donor and recipient ewes.

TABLE 10.2: Sequence of Events in an Ova Transfer Programme

Days from start of programme	Donors	Recipients
0	Insert intravaginal sponges	
14	Remove intravaginal sponges	
16	Oestrus	Insert intravaginal sponges
29	Inject PMSG	
30		Remove intravaginal sponges
30	Oestrus and insemination	Oestrus
35	Recovery of ova	Transfer of ova

The dose of PMSG should be chosen to achieve the highest ovulation rate without any of the problems of overstimulation. Fertilisation and egg recovery should be satisfactory with a mean ovulation rate in the range of 8-12. Ova are usually recovered 3-3½ days after mating (8-16 cell-stage) and best survival rates are achieved by transfer direct into the uterus. A thin polythene tube is inserted into the Fallopian tube, firmly held by an assistant and directed into a small glass dish. Each Fallopian tube and 3-4cm of uterus is flushed by inserting a blunt needle into the uterine lumen and flushing back through the utero-tubal junction with about 5ml of sterile sheep serum at 37°C.

Ova may also be recovered from the uterus 5 days or later after mating. This technique has the advantage of causing less adhesions on the reproductive tract than with tubal flushing and the ova recovery rates are satisfactory. A catheter with an inflatable cuff is inserted into the uterine horn. Following inflation of the cuff a blunt needle is introduced into the uterus in the region where the Fallopian tubes enter and serum is flushed through the uterus and the catheter and into collecting dishes.

Cleavage is used as the criterion of fertilisation and ova for transfer are sucked up in a fine glass pipette with a minimum of fluid. These are gently aspirated into the uterus of the recipient ewe through a small puncture in the uterine wall. Ova should be maintained at 35-37°C between the time of recovery and transfer. Ova recovered at the 2 or 4 cell-stages should be transferred into the anterior end of the Fallopian tubes for best results.

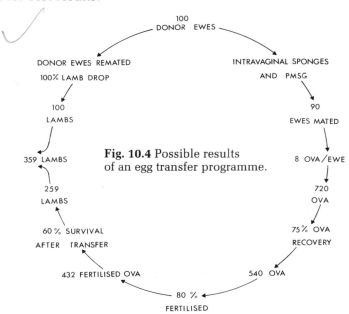

Fig. 10.4 Possible results of an egg transfer programme.

Possible Results of an Ova Transfer Programme

Fig. 10.4 outlines the possible outcome of a programme where ewes such as Romney are treated with 1500 iu PMSG and ova transfers attempted. Ovum transplantation involves a series of procedures and varying levels of success at any stage can affect the final result. Provided treatment of ewes commences after the start of the breeding season the number of ewes not mated should be less than 10%. The ovarian response in individual ewes is likely to be extremely variable. If rams are producing high quality semen and ewes in oestrus are mated several times fertilisation rates should be 80% or higher. The percentage of ova surviving after transfer of one or two 8-16 celled ova is also variable but should be 50-70%. The percentage surviving after transfers of single ova is often a little greater than when two ova are transferred. However, if a large number of ova are recovered this means twice as many recipient ewes will need to be used. Consequently two ova are often transferred to each recipient ewe.

Problems of Ova Transfer

In ewes which undergo more than one donor operation the response to a constant dose of PMSG may decrease and with repeated stimulations some animals will fail to ovulate. Whether this is due to an immune response to the injected foreign protein or due to a temporary exhaustion of ovarian follicles at a suitable stage of development, or both, is not clear.

Lowered fertilisation rates and egg recovery rates, possibly due in part to rapid ova transport down the Fallopian tubes, puts an upper limit on the level of gonadotrophic stimulation possible. Adhesions of the reproductive tract are also a problem frequently encountered if ewes are subjected to surgical recovery of ova more than once and these may, in serious cases, make the ewe infertile. It is essential during surgery to minimise handling of the reproductive tract and also to minimise the amount of blood which comes in contact with the tract. With careful procedures two or three ova recovery operations on the same ewes should be possible without major problems due to adhesions.

Costs

The cost of hormonal treatment for donor and associated recipient ewes will depend on the type of synchronising treatment used and the type and availability of PMSG. Insertion of intravaginal sponges for synchronisation and injection of donors can easily be carried out by non-skilled personnel. However the recovery and transfer of ova must be conducted by highly skilled operators. The costs of such labour together with the provision of surgical and microscopic equipment, and materials must be computed for the enterprise. There will also be an opportunity cost for loss of production from recipient ewes to be included.

ARTIFICIAL INSEMINATION (AI)

AI in sheep has been extensively reviewed.[48] The main advantage of AI is that it allows widespread use of superior sires. Other commonly cited advantages are a reduction in ram costs and the use of old or incapacitated rams. In assessing the value of this technique for flock improvement the dangers of using insufficiently proven rams as opposed to an extended generation interval, if sires are fully proven, must be considered as well as the dangers of inbreeding and the spreading of undesirable recessive defects.

The organisation of an AI programme requires detailed planning and involves practical skills in sheep handling and insemination procedures. Considerable costs are likely to be incurred because of the labour required with AI. It is often convenient to synchronise oestrus in ewes prior to AI to reduce the insemination period. Ewes are commonly inseminated at the second oestrus following progestagen treatment at which time fertility is normal.

Regardless of possible genetic gains, and therefore presumably gains in production, the monetary advantage accruing from an AI programme must at least balance the added cost. The considerable sexual capacity of rams,[3 5] which is not generally used in industry, should also be considered. If 200-300 ewes were to be mated by one ram it would be advisable to use natural mating thus incurring no costs for equipment, drugs and labour. Also conception rates following AI are generally lower than those following natural mating.

Semen Collection and Evaluation

Two methods are used to induce ejaculation in rams:
(1) *Electro-ejaculation* (EE) uses low voltage pulses transmitted by a probe in the rectum;
(2) *Artificial vagina* (AV).

Ejaculates collected by AV are more dense but often of lower volume than those from EE. There is also the risk of urine contamination with EE and some rams are, or become, refractory to this method of collection. Training of rams for AV is not difficult and most rams will serve within a short period. Collection by AV should therefore be preferred in organisation of an AI programme. Following collection, semen should be maintained at 30°C. Care should be taken to avoid contact with water and disinfectants and the semen should be kept out of direct sunlight. If possible inseminations should normally be carried out within one hour of collection.

Assessment of the ejaculate can be made using density and motility of sperm estimations. The consistency and colour of ejaculates gives a good approximation of semen density and can be graded (0 to 5 scale).

253

A thick creamy appearance (grade 5) indicates 4.5 to 5.5 x 10^9 spermatozoa per ml whereas a thin creamy appearance (grade 3) indicates 2.5 to 3.5 x 10^9 spermatozoa per ml. Density can be more accurately determined by haemocytometer or colorimetric methods. Semen motility, so critical for the transport of spermatozoa within the reproductive tract, is also usually assessed on a 0-5 scale. For AI it is advisable to use only semen graded 4/4 or better for density and motility. Microscopic examination of stained semen smears also allows estimation of the percentages of abnormal and live sperm cells. These latter criteria are both correlated with wave motion score which is a much more simple and useful criterion for semen evaluation in the field.

Insemination

In two or three ejaculates of semen most rams will produce at least 6 x 10^9 spermatozoa each day during an AI programme. With 120 to 125 x 10^6 spermatozoa in a volume of 0.1ml being necessary for maximum fertility,[47] this amount of semen would be sufficient to inseminate about 50 ewes. Collection of more ejaculates on specific days or more spermatozoa per ejaculate from some rams often makes it possible to inseminate much higher numbers of ewes. Density of the inseminate is more important than volume in determining fertility,[10] this being more critical in ewes inseminated at a synchronised heat when the transport of spermatozoa within the reproductive tract is impaired. With ewes inseminated at synchronised oestrus, undiluted semen should be used. If dilution is necessary this should not be more than 1:1. There is little point in increasing numbers of spermatozoa by increasing the volume of the inseminate above 0.1ml. Dilution of semen is usually with egg yolk-glucose-citrate or with heat-treated cows' milk. The rate of dilution will depend on the density of the collected ejaculates.

Restraint of ewes for AI must be as gentle as possible as stress will impair the transport of spermatozoa. The inseminate is introduced 1-2cm into the entrance of the cervix. Best results are achieved when ewes are inseminated approximately 12 hours after the onset of oestrus. A further insemination 10-12 hours later will result in a 6-10% increase in fertility.[48]

Storage of Semen

Storage of ram semen would be of real advantage in AI programmes.

Semen can be chilled and stored at 2-5°C for up to 24 hours after dilution with a milk or egg yolk extender. In progestagen-synchronised ewes, where dense inseminates are required for high fertility, dilution rates must be low but the higher the concentration of sperm the lower is the survival rate during storage.[18] Whatever the dilution rate, sperm survival decreases with time and best fertility will occur when storage is for only a few hours. Short-term storage has only limited application in industry.

Semen may be stored at -196°C for long periods. Tris-based diluents containing egg yolk, citric acid and glycerol (protective against cold shock) have been widely used to dilute the semen which is then gradually cooled to 5°C in 1.5 - 2 hours. Freezing is carried out in straws, containing 0.25 - 0.50ml, suspended in liquid nitrogen vapour or by pelleting (0.1 - 0.4ml) on dry ice at -79°C. Thawing of pellets or straws is at 37°C. If a thawing solution is used it has the disadvantage of further diluting the thawed semen which may need centrifugation to reconcentrate the sperm before insemination. This can be avoided when the pre-freezing dilution does not exceed 1:2 and no thawing solution is used. Semen frozen in straws can not be easily reconcentrated.

Fertility following insemination of previously frozen sperm has been variable and below that of fresh undiluted or diluted semen.[48] This is due to lower motility of the thawed inseminates. By-passing of the cervix by surgical insemination direct into the uterus results in high fertilisation rates but may also cause a high level of embryonic mortality. Highly variable results make AI with frozen semen a doubtful commercial proposition until high ewe fertility is possible on a repeatable and extensive scale.

EFFICIENCY OF RAM USAGE

Ram/Ewe Ratios

The use of fewer but more productive sires over large numbers of ewes is perhaps the easiest method of flock improvement available. The ram/ewe ratios used in industry generally underestimate the sexual capacity of rams. Ratios of 50 ewes per ram are frequently used. A number of trials have shown that for flock mating a ram/ewe ratio of 1/100 is sufficient and rams may be successfully joined with higher numbers of ewes. When ram/ewe ratios from 1/50 to 1/210 have been used, within experiments there have been no large differences between the percentage of ewes mated in the first cycle, returns to service, barrenness or twinning.[3] Studies of semen quality from the rams used in these experiments showed a massive decline in volume, density and number of sperm per ejaculate a short time after the start of mating but little difference between groups of rams joined with different numbers of ewes.

The results of field trials suggest that at least double the usual "farmer number" of ewes can be joined successfully with groups of rams without a loss of fertility as shown in Table 10.3.

TABLE 10.3: Mating Performance of Individual Rams and Returns to Service in Groups of Rams joined with Different Numbers of Ewes.[3]

Ram/ewe	Total ewes mated-first 17 days	Number of ewes mated by each ram in first 17 days			% returning to service
3/150	136	122	112	96	14.1
3/450	413	297	268	205	16.3
3/210	171	144	117	111	14.6
3/630	520	365	338	332	12.1

The number of ewes mated by each ram of a group (Table 10.3) can be determined by crayon marks but the number of services performed by each ram is likely to be well in excess of these numbers. Many oestrous ewes are mated more than once by the same ram.[5] When most rams are capable of mating 200 or more ewes in a 17 day period conventional ram/ewe ratios must be questioned. This particularly applies with group mating, as most ewes will be mated by more than one ram and any infertile or inactive male will be compensated for by others in the group. The use of fewer but more productive rams to cover more ewes certainly has greater farmer application than use of AI.

Mating Performance of Rams

Many inter-related factors such as ages of rams and ewes, paddock size, nutrition, season of the year, competition between rams and previous exposure of rams to ewes will affect the mating performance. Of these competition is unlikely to be of importance for paddock mating.[22]

The duration of oestrus may be shorter in 1½ year-old ewes than in mature ewes although this view is equivocal. If "fresh" rams are introduced every few hours, thus eliminating mating fatigue and competition between the 1½ year ewes and older animals, then estimates of the duration of oestrus are similar for the two age-groups.[7] However, ram-seeking activity is lower in 1½ year ewes and also in ewes of lower live weight within a flock. When 1½ year ewes are joined in conjunction with older ewes they are less competitive for the ram's attention. This is more evident when the number of ewes per ram is increased.[4] Consequently it is advisable to join mobs of 1½ year ewes separately with a lower number of ewes per ram than the older ewes.

The effect of age of the ram on mating efficiency is not clear. Lower flock fertility following mating with 1½ year rams[32] is due to a lower number of ewes mated[22] or to reduced semen quality.[51] In 1½ year rams, not previously joined with ewes, failure to mate has sometimes been reported in Australia[34] but this does not seem to be a problem in

New Zealand conditions. Young rams have been shown to have comparable mating activity to older rams at a ram/ewe ratio of 1/60, but to have a somewhat lower activity at 1/180.[5] This reduction in activity appears to have few detrimental effects on ewe fertility.

There is little experimental information concerning sexual activity and fertility of ram lambs used at 7-8 months of age. Ram lambs which have been reared under good nutritional conditions will reach puberty at an earlier age and a higher live weight than when lower levels of feeding are used and can be successfully used to mate ewes. Ram lambs are often used in New Zealand with very satisfactory results[36] although they may be less effective than 1½ year rams in detecting oestrus in synchronised ewes and have lower fertility.[16] Farmer experience with mating of ram lambs indicates that ram/ewe ratios of 1/30 to 1/50 can be confidently recommended.[36] This will allow rams to be used at an earlier age provided their breeding value can be predicted at this stage or if progeny-test information is to be collected. The importance of paddock size on the probability of ram/oestrous ewe contact is not clear as spatial arrangement of sheep may vary according to the breed of ewe, or the shape and topography of a paddock. An increase in paddock size has been shown to be detrimental to the fertility of 1½ year ewes and will cause a substantial decrease in the percentage of ewes which will migrate to tethered rams and be mated by them.[7] It seems that shape, topography and ground cover of mating areas may be more important than the absolute area.[6] Ewes should not be able to become divided from rams by physical barriers such as creeks, deep gullies and areas of thick bush or scrub within a mating paddock. As most ewes will be mated and conceive within a short time after joining, attention should be given to selection of the most suitable mating paddocks particularly for the first month of the mating period.

Effects of nutrition on libido and semen quality will also affect efficiency of ram usage. Undernutrition will depress ram libido within 5-10 weeks.[33] The energy content of the diet appears more critical than protein content in affecting spermatogenesis and therefore semen quality. These effects are probably due to effects on testis size as differences in nutrition will affect this parameter and each gram of testicular tissue produces about 20×10^6 spermatozoa, this remaining constant over a wide variety of nutritional conditions. Caution is required in interpretation of such information for industry use. There is little doubt that rams which have been subjected to high levels of feeding for long periods and are over-fat have reduced sexual activity. Probably under-nutrition would need to cause substantial decreases in live weight before decreasing libido and semen quality sufficiently to affect flock fertility.

Libido and semen production of rams in New Zealand are maximal during the breeding season (March to June). There are declines in both characteristics in the non-breeding season in most environments.

257

Rearing of rams in homosexual groups has been suggested as a reason for high levels of sexual inhibition in 1½ year-old rams.[34] There is however no conclusive evidence that this may affect subsequent patterns of mating behaviour. Certainly in New Zealand flocks where sexual inhibition of young rams does not seem to be a problem, exposure to ewes before first mating seems unimportant.

Single Sire Versus Group Mating Systems

Factors affecting ram fertility become more critical with single-sire mating than when groups of rams are used. In single-sire mating systems the numbers of ewes mated, returns to service and barrenness reflect the fertility of individual rams; in a group mating system relatively infertile or non-active rams tend to have little effect on flock fertility, as many ewes are mated by more than one ram.

Returns to service are usually 7-10% higher in ewes mated by one ram than in ewes mated by two or more rams[3] although in conditions where ewe and ram fertility is low there may be greater advantages in increasing the number of services or number of rams mating each ewe. Levels of barrenness in stud Romney flocks, where single-sire mating is used, are about 7% in comparison with about 4% in commercial Romney flocks where group mating is used.[39] [40] For the stud flocks the estimates may be somewhat reduced due to the possibility that mating groups are sometimes combined towards the end of the mating period with additional rams.

Within single-sire mating systems groups of ewes may either be joined in paddocks or following detection of oestrus with vasectomised rams, be transferred to rams in small pens. In some situations this may not affect ewe fertility[17] but pen mating will often result in a substantial decrease in the percentage of ewes mated.

Testing Mating Performance of Rams

Rams can be assessed for semen quality and libido but the tests are often not well related to performance during natural mating. Collection of a dense motile ejaculate by electro-stimulation is sufficient to predict a ram will be fertile, but a sample of poor density and motility or complete failure to collect does not invariably mean a ram will be infertile. Rams should therefore be retested before rejection. The use of EE for fertility assessments during mating will often result in failure to collect any sample although ram fertility is still satisfactory. This is due to depletion of epididymal sperm reserves and low volume and density of ejaculates in working rams.

Libido tests consist of joining rams with oestrous ewes, usually in a small pen for a short period, and observing the number of mounts and services. Determination of subsequent flock mating activity on this basis could be useful in allocating different numbers of ewes to join with particular rams. Providing rams mate at least one ewe in a series of libido tests there seems to be no relationship with this performance and flock mating activity.[29]

Libido tests can also be of use in identifying "non worker" rams. However with further testing and especially among young puberal animals as they develop, many will show mating activity, but it is difficult to predict when this will occur.

CONTROL OF PARTURITION

Parturition in sheep is initiated by the secretion of glucocorticoid hormones (mainly cortisol) from the foetal adrenal gland. Artificial stimulation of this phenomenon by administration of glucocorticoid in late stages of gestation in ewes will induce premature parturition. Thus the intramuscular injection of 8 or 16 mg of dexamethazone on days 142-144 of gestation will induce parturition in most ewes approximately two days later[14] and concentrate lambing in comparison with untreated animals. Treatment before 140 days of gestation is often effective in inducing parturition but lamb mortality may be increased. Injection of oestrogen has been shown to be at least as effective as dexamethazone in the induction of parturition[42] and a single dose of 10-40 mg of ODB between days 140 and 144 of gestation resulted in 14%, 79% and 6% of ewes lambing on the first, second or third days after treatment, respectively. Thus the use of dexamethazone or oestrogen can be used to synchronise lambing. However, uses of the technique within industry will remain limited to a few specialised situations.

REDUCTION OF LAMBING LOSS

Within New Zealand flocks 5-25% of lambs born die between birth and weaning. The survival rate of lambs is lower in 1½ year-old ewes than in older ewes and is higher in single-lambs than in multiple-born lambs. Survival is also affected by birth weight, sex of the lamb and disease. Periods of wet windy and cold weather cause lamb losses to increase markedly.

Nutrition, Birth Weight and Lamb Survival

The chance of lambs surviving is related to birth weight and therefore to nutrition in late pregnancy. Below and above a certain range in birth weights lamb losses among singles and twins increase. For Romney

259

and Border Leicester x Romney lambs these ranges are about 3.2 to 4.5 kg for twins and 3.8 to 5.0 kg for singles. Low levels of nutrition in late pregnancy will have less of an effect on the birth weights of single- than on twin-born lambs. Similarly very high levels of nutrition throughout pregnancy will cause high birth weights and often an increase in the incidence of dystocia particularly in single-bearing ewes. This is particularly important with 1½ year ewes. Also in ewe hoggets, which lamb about a month after the main flock, most will have only one lamb and the plentiful spring pasture supply requires that intake should be restricted to keep birth weights of lambs low to limit the incidence of dystocia. The use of sires of smaller mature size (e.g. Southdown and Cheviot) will also result in dystocia being less of a problem.

Provision of Shelter and Intensity of Shepherding

Many lamb losses due to starvation and exposure soon after birth can be avoided by the provision of shelter in the paddocks or by lambing the ewes indoors. In a study in Australia deaths were 21% of singles and 58% of twins from ewes lambing in paddocks completely exposed to prevailing weather in comparison with 8% and 24% in ewes lambed under cover. Most of the losses occurred in a period of very inclement weather. Subsequent experiments have also shown substantial reductions in lamb mortality in sheltered compared with exposed flocks.[38] Provision of shelter belts (trees) in lambing paddocks exposed to prevailing winds would seem a cheaper alternative than building of expensive sheds in which to lamb ewes. The correct siting of shelter belts in exposed paddocks is likely to be of importance, yet the normal behaviour of the ewe close to lambing may still isolate her from the remainder of the sheltering flock. Protective plastic covers put on lambs shortly after birth should also keep lambs warmer and reduce losses. These are widely used especially during storms but can cause problems of ewes failing to accept lambs, particularly one of a pair of twins. Disturbance of ewes soon after lambing is also suspected to contribute to mis-mothering and eventual starvation of the young lamb.

Increasing interest exists in "easy care" sheep as a means to contain farm labour costs. At lambing time this means that ewes should lamb unassisted and should also rear their complement of lambs without any assistance. Adoption of a minimal or no-shepherding system necessitates that levels of dystocia are kept low if ewe deaths are not to rise. Some comparisons of twice daily shepherding versus once every two days have shown minimal differences between the two systems. Usually more ewes will be assisted at lambing in an intensively shepherded group and deaths among twin lambs will be lower. Perhaps more frequent shepherding is necessary in adverse weather conditions particularly for mothering up multiple-born lambs, but it seems that much more shepherding is done in industry than is necessary. Provision of shelter, judicious feeding in late pregnancy and selection of sires with high survival rates in their progeny can all contribute to the success of an easy care philosophy at lambing time.

REFERENCES

[1] Allison, A. J. 1968: The influence of liveweight on ovulation rate in the ewe. *Proc. N.Z. Soc. Anim. Prod., 28:* 115.

[2] Allison, A. J. 1975a: Effect of nutritionally induced live weight differences on the ovarian response and fertility in ewes treated with pregnant mare's serum gonadotrophin. *N.Z. Jl. agric. Res., 18:* 101.

[3] Allison, A. J. 1975b: Flock mating in sheep 1. Effect of number of ewes joined per ram on mating behaviour and fertility. *N.Z. Jl. agric. Res. 18:* 1.

[4] Allison, A. J. 1977: Flock mating in sheep 2. Effect of number of ewes joined per ram on mating behaviour and fertility of 2-tooth and mixed-age ewes run together. *N.Z. Jl. agric. Res. 20:* 123.

[5] Allison, A. J. 1978a: Flock mating in sheep 3. Comparisons of two-tooth and six-tooth rams joining with differing numbers of ewes per ram. *N.Z. Jl. agric. Res. 21:* 113

[6] Allison, A. J.; Davis, G. H. 1976a: Studies of mating behaviour and fertility of Merino ewes. I. Effects of number of ewes joined per ram, age of ewes and paddock size. *N.Z. Jl Exp. Agric. 4:* 259

[7] Allison, A. J.; Davis, G. H. 1976b: Studies of mating behaviour and fertility of Merino ewes. II. Effects of age, live weight and paddock size on duration of oestrus and ram-seeking activity. *N.Z. Jl Exp. Agric. 4:* 269. 269.

[8] Allison, A. J.; Kelly, R. W. 1978: Synchronisation of oestrus and fertility in sheep treated with progestagen impregnated implants and prostaglandins with or without intravaginal sponges and subcutaneous pregnant mares' serum. *N.Z. Jl agric. Res. 21:* 389.

[9] Allison, A. J.; Robinson, T. J. 1970: The effect of dose level of intravaginal progestagen on sperm transport, fertilisation and lambing in the cyclic Merino ewe. *J. Reprod. Fert. 22:* 55.

[10] Allison, A. J.; Robinson, T. J. 1971: Fertility of progestagen treated ewes in relation to the numbers and concentration of spermatozoa in the inseminate. *Aust. J. biol. Sci. 24:* 1001.

[11] Allison, A. J.; Stevenson, J. R.; Kelly, R. W. 1977: Reproductive performance and wool production of Merino and high fertility (Booroola) cross Merino ewes. *Proc. N.Z. Soc. Anim. Prod. 37:* 230.

[12] Averill, R. L. W. 1964: Ovulatory activity in mature Romney ewes in New Zealand. *N.Z. Jl. agric. Res. 7:* 514.

[13] Bindon, B. M.; Ch'ang, T. S.; Turner, H. N. 1971: Ovarian response to gonadotrophin by Merino ewes selected for fecundity. *Aust. J. agric. Res. 22:* 809.

[14] Bosc, M. J. 1972: The induction and synchronisation of lambing with the aid of Dexamethasone. *J. Reprod. Fert. 28:* 347.

[15] Ch'ang, T. S. 1961; Reproductive performance of New Zealand Romney sheep grazed on red clover *(Trifolium pratense)* pastures. *J. agric. Sci., Camb., 57:* 123.

[16] Clarke, J. N.; Roberts, E. M.; Carter, A. H.; Kirton, A. H. 1966: Hormonal synchronisation of oestrus in Romney ewes during the breeding season. *Proc. N.Z. Soc. Anim. Prod. 26:* 107.

[17] Clarke, J. N.; Geenty, K. G.; Bennett, R. G.; Christensen, G. N.; Wilson, J. A. 1974: Comparison of pen and paddock systems for the pedigree mating of sheep. *Proc. N.Z. Soc. Anim. Prod. 34*: 23.

[18] Colas, G.; Courot, M. 1976: Storage of ram semen: *In Sheep Breeding*. (Eds G. J. Tomes, D. E. Robertson, and R. J. Lightfoot.) Western Australian Institute of Technology, Perth. p. 455.

[19] Coop, I. E. 1962: Live weight-productivity relationships in sheep I. Live weight and reproduction. *N.Z. Jl. agric. Res. 5*: 249.

[20] Coop, I. E. 1966: Effect of flushing on reproductive performance of ewes. *J. agric. Sci., Camb. 67*: 305.

[21] Coop, I. E.; Clark, V. R. 1960: The reproductive performance of ewes mated on lucerne. *N.Z. Jl. agric. Res. 3*: 922.

[22] Croker, K. P.; Lindsay, D. R. 1972: A study of the mating behaviour of rams when joined at different proportions. *Aust. J. Exp. Agric. Anim. Husb. 12*: 13.

[23] Dyrmundsson, O. R. 1973: Puberty and early reproductive performance in sheep I. Ewe lambs. *Anim. Breed. Abstr. 1*: 273.

[24] Edgar, D. G.; Bilkey, D. A. 1963: The influence of rams on the onset of the breeding season in ewes. *Proc. N.Z. Soc. Anim. Prod. 23*: 79.

[25] Gordon, I. 1958: The use of progesterone and serum gonadotrophin (PMSG) in the control of fertility in sheep II. Studies in the extra-seasonal production of lambs. *J. agric. Sci., Camb. 50*: 152.

[26] Hight, G. K.; Jury, K. E. 1969: Lamb mortality in hill country flocks. *Proc. N.Z. Soc. Anim. Prod. 29*: 219.

[27] Hunter, G. L. 1968: Increasing the frequency of pregnancy in sheep. 1. Some factors affecting rebreeding during the post-partum period. *Anim. Breed Abstr. 36*: 347. 2. Artificial control of rebreeding and problems of conception and maintenance of pregnancy during the post-partum period. *Anim. Breed Abstr. 36*: 553.

[28] Joyce, M. J. B. 1972: A comparison of three mating systems in ewes treated with intravaginal (SC - 9880/progesterone) progestagen and PMS. *VII Int. Congr. Anim. Reprod. Artif. Insem., Munich 2*: 935.

[29] Kelly, R. W.; Allison, A. J.; Shackell, G. H. 1975: Libido testing and subsequent mating performance in rams. *Proc. N.Z. Soc. Anim. Prod. 35*: 204.

[30] Knight, T. W. 1973: A study of factors which affect the potential fertility of the ram. *Ph. D Thesis, Univ. of Western Australia.*

[31] Lightfoot, R. J. 1968: Studies on the number of ewes joined per ram for flock matings under paddock conditions. 2. The effect of matings on semen characteristics. *Aust. J. agric. Res. 19*: 1043.

[32] Lightfoot, R. J.; Smith, J. A. C. 1968: Studies on the number of ewes joined per ram for flock matings under paddock conditions. 1. Mating behaviour and fertility. *Aust. J. agric. Res. 19*: 1029.

[33] Mattner, P. E.; Braden, A. W. H.; Turnbull, K. E. 1967: Studies in flock mating of sheep. 1. Mating behaviour. *Aust. J. Exp. Agric. Anim. Husb. 7*: 103.

[34] Mattner, P. E.; Voglmayr, J. K. 1962: A comparison of ram semen collected by the artificial vagina and by electro-ejaculation. *Aust. J. Exp. Agric. Anim. Husb. 2*: 78.

[35] Mauleon, P. 1976: Manipulation of the breeding cycle. In *Sheep Breeding*. (Eds. G. J. Tomes; D. E. Robertson; R. J. Lightfoot), Western Australian Institute of Technology, Perth. 310.

[36] McDonald, M. F. 1974: The use of ram lambs as sires. *Sheepfmg A.*: 1.

[37] McDonald, M. F.; Ch'ang, T. S. 1966: Variation in ovarian activity of Romney Marsh ewes. *Proc. N.Z. Soc. Anim. Prod. 26*: 98.

[38] McLaughlin, J. W.; Egan, J. K.; Poynton, W. McL.; Thompson, R. L. 1970: The effect upon neonatal lamb mortality of lambing systems incorporating the use of partial and complete shelter. *Proc. Aust. Soc. Anim. Prod.* 8: 337.

[39] Quinlivan, T. D.; Martin, C. A. 1971a. Survey observations on the reproductive performance of both Romney stud and commercial flocks throughout New Zealand. 2. Lambing data from an intensive survey in stud flocks. *N.Z. Jl. agric. Res.* 14: 858.

[40] Quinlivan, T. D.; Martin, C. A. 1971b: Survey observations on the reproductive performance of both Romney stud and commercial flocks throughout New Zealand. 3. National commercial flock performance. *N.Z. Jl. agric. Res.* 14: 880.

[41] Rattray, P. V.; Jagusch, K. T.; Smith, J. F.; Winn, G. W.; MacLean, K. S. 1980: Flushing responses from heavy and light ewes. *Proc. N.Z. Soc. Anim. Prod.* 40: 34.

[42] Restall, B.; Herdegen, F.; Carberry, P. 1976: Induction of parturition in sheep using oestradiol benzoate. *Aust. J. Exp. Agric. Anim. Husb.* 16: 462.

[43] Robinson, T. J. 1951: The control of fertility in sheep Part II. The augmentation of fertility by gonadotrophin treatment of the ewe in the normal breeding season. *J. agric. Sci., Camb.,* 41: 6.

[44] Robinson, T. J. 1964: Synchronisation of oestrus in sheep by intravaginal and subcutaneous applications of progestin impregnated sponges. *Proc. Aust. Soc. Anim. Prod.* 5: 47.

[45] Robinson, T. J. (Ed.) 1967: *The Control of the Ovarian Cycle in Sheep.* Sydney Univ. Press.

[46] Robinson, T. J. 1970: Fertility following synchronisation of oestrus in the sheep with intravaginal sponges. 2. Effect of dose of progestagen and rate of absorption. *Aust. J. agric. Res.* 21: 783.

[47] Salamon, S. 1962: Studies on the artificial insemination of Merino sheep. 3. The effect of frequent ejaculation on semen characteristics and fertilising capacity. *Aust. J. agric. Res.* 13: 1137.

[48] Salamon, S. 1976: *Artificial Insemination of Sheep.* Sydney Univ. Press.

[49] Tervit, H. R.; Havik, P. G.; Smith, J. F. 1977: Effect of breed of ram on the onset of the breeding season in Romney ewes. *Proc. N.Z. Soc. Anim. Prod.* 37: 142.

[50] Wallace, L. R. 1954: Studies in the augmentation of fertility of Romney ewes with Pregnant-Mare Serum. *J. agric. Sci., Camb.,* 45: 60.

[51] Wallace, L. R. 1961: Influence of live weight and condition on ewe fertility. *Proc. Ruakura Fmrs. Conf.*: 2.

[52] Wallace, L. R.; Lambourne, L. J.; Sinclair, D. P. 1954: Effect of pregnant mare serum on the reproductive performance of Romney ewes. *N.Z. J. Sci. Tech., A.* 35: 421.

Index